Understanding Ultrasonic Level Measurement

Stephen Milligan, B.Sc.
Henry Vandelinde, Ph.D.
and Michael Cavanagh

MP MOMENTUM PRESS

MOMENTUM PRESS, LLC, NEW YORK

Published by:

Momentum Press®, LLC
222 East 46th Street
New York, NY 10017
www.momentumpress.net

ISBN-13: 978-1-60650-439-0 (hardcover, casebound)
ISBN-10: 1-60650-439-8 (hardcover, casebound)
ISBN-13: 978-1-60650-441-3 (e-book)
ISBN-10: 1-60650-441-X (e-book)
DOI: 10.5643/9781606504413

Cover design by Jonathan Pennell

10 9 8 7 6 5 4 3 2 1

Printed in the United States of America

Contents

Chapter Five
Installation   61

Acknowledgements

As you can imagine, a project like this involves the efforts and contributions of many people. To begin with, the authors want to thank the generations of engineers, designers, application specialists, sales people, support staff, and management who have developed the technology and the products over the years. All of us also owe a huge debt of gratitude to our customers who have allowed us to grow and to share in their successes by participating in our vision. All together, they have created the SITRANS LUT400, the revolutionary ultrasonic controller with one millimeter accuracy the markets have been waiting for.

The authors also want to thank all of the writers and photographers who have contributed material used in this book, both in specific content and for general background information. They are too numerous to mention, but their enthusiasm for the technology and their efforts are much valued. The artistic contributions of Peter Froggatt are also appreciated. Over the years, his drawings and photos have helped define the product line, and his work graces many of the pages in this humble tome. Those who took the time to edit and provide comments and other input also have our gratitude.

Specifically, we want to thank the editing and organizational skills of Jamie Chepeka. Her dedication to the project was unwavering, even in the face of looming deadlines and creative angst. Without her management guidance, we would still be staring at our screens.

Lastly, the authors apologize in advance for any and all mistakes, inaccuracies, and omissions. We take full responsibility and assure you that we will do better next time.

Chapter One

History of ultrasonics

How sweet that joyous sound,/ whenever we meet.[1]

Siemens Milltronics Process Instruments has a long and successful history specializing in the manufacture of equipment for industrial process measurement. Based in Peterborough, Canada, Siemens Milltronics (PI2) is now a key member of the Sensors and Communication division within the Siemens Industry division, supplying instrumentation across the globe.

Founded in 1954 by Stuart Daniel, a former employee of Canadian General Electric, the company began as Milltronics and engineered electronic ball mill grinding controls for the cement and mining industry. From this, the company expanded and diversified its product line to develop a wide range of process measurement devices. It has become a leader in level measurement technology. The Siemens Milltronics range of instrumentation now includes ultrasonic, radar, and capacitance technologies, but the foundation of its innovation and successful design and technical expertise lies in its ultrasonic echo-ranging technology.

Siemens Milltronics ultrasonic echo-ranging technology comprises highly sophisticated instrumentation applying digital circuitry to ultrasonic echo-ranging. This innovation has produced a range of technologically advanced products capable of monitoring liquid and solids levels from a few centimeters to over 60 meters (200 ft). To date, over 1,000,000 points of level on a diverse range of material, including solids, liquids, slurries, and resins, are monitored across the globe by Siemens Milltronics, many in hostile and hazardous environments.

The Siemens Milltronics ultrasonic product line is constantly improving as technological advances are implemented, new products are

---

[1] Van Morrison, "Joyous Sound." *A Period of Transition,* 1977.

developed, and new applications are tackled and won over. Complemented by a team of highly skilled applications engineers, service personnel, and a dedicated Siemens sales force, Siemens Milltronics continues to provide reliable and innovative level solutions to industry across the globe.

Ultrasonics and level measurement

The measurement of level has been integral to human development since pre-industrial times.

"Egypt," Herodotus remarked more than 2000 years ago, referring to the vast irrigation project that sustains that country's agriculture, "is the gift of the river." Every June, as snowmelts from the Tanzanian Highlands and spring rain from the Congo begin accumulating in the Nile, its elevation begins to rise. It rises gently to a crest in late September or early October, then subsides by late December. Seed goes into the rich, freshly deposited silt as soon as the flood recedes.

Egyptian engineers began capturing the river for irrigation projects about 7,000 years ago. Because the system relies on a complicated system of gates to distribute water across a broad, relatively flat area, it's vital that engineers know the height of the river in advance of its arrival. The first solution was to simply mark the riverbanks and convey information back to headquarters via runners. Later, engineers developed a large variety of "nilometers," devices used to measure the river height. Most, however, consisted of ordinary graduated scales that projected vertically upward from the riverbed and were read directly.

Today, the U.S. Geological Survey and the National Oceanic and Atmospheric Administration use similar devices: graduated poles stuck into the water. Technicians read most of them manually, but there are some in flood-prone areas that transmit information directly to the agency via radio. Though millennia-old solutions for measuring river level are still in use, there are thousands of level-determination problems in industry that demand much more sophisticated solutions. Like their forebears, contemporary engineers have responded with impressive ingenuity.[2]

---

[2] Felton, Bob. "Level Measurement: Ancient Chore, Modern Tools." ISA, August 2001.

Ingenuity is also the key to the success of Siemens Milltronics ultrasonic technology as it meets the demands of level measurement in the process systems market. The need for process measurement dates back to the Industrial Revolution when the development of the steam engine created a requirement for the accurate measurement of temperature, pressure, and flow.

By the early twentieth century, process engineers were determining process measurements using a variety of mechanical devices including floats, sight glasses, thermometers, gauges, and armatures. Accuracy was often elusive, and these devices were supplemented by human experience. Process engineers often relied on their senses to complement the technology: using sight, sound, touch, smell, and even texture, engineers would examine process smoke, liquid clarity, texture, and smell to determine product quality. However, chemical compounds, safety restrictions, system complexity, and awareness now make this type of tactile verification impossible, requiring measurement to be made by the instrument alone.

Process measurement incorporates a variety of solutions, from pressure and temperature to flow and level. While Siemens SC PI offers instrumentation to measure all of these, Siemens Milltronics specializes in the calculation of level.

Level measurement instrumentation currently employs a variety of sophisticated technologies, with ultrasonic measurement as the cornerstone. The origins of ultrasonic measurement technology lie in early use by submarines of sonar for depth gauging and marine detection, but it wasn't until 1949 that these principles were applied to level measurement. Bob Redding, of Evershed and Vignoles, developed an ultrasonic instrument with servocontrol that automatically measured oil level and then transferred that information to a remote indicator.

Other technologies were also applied to remote level measurement by companies like Magnetrol, which applied its magnetic switching technology to the control of pumps and other devices for use in water level alarming. The device transmitted level changes to the switch mechanism without any mechanical or electrical connection and eliminated mechanical devices such as diaphragms and stuffing boxes.

In 1963, Magnetrol introduced Modulevel©, the first magnetically coupled pneumatic proportional level control. The first significant

© Modulevel is a registered trademark of Magnetrol.

sensing instrument, it led the way to new markets in continuous process level control. By the 1970s, ultrasonic technology, already used in ship and plane detection, was developed for the measurement industries. Sonar principles were applied to use in air, using modified low frequency sonar equipment with piezoelectric crystals to generate echo ranging. These new sensors were applied to process control tasks such as point level, continuous level, concentration, and full pipe applications. In the mid-1980s, analog instrumentation went digital and offered 4 to 20 mA signal, opening up communication possibilities, and greatly increasing its value as control instrumentation.

Milltronics entered the market in these early days of ultrasonic development. In 1973, after being the main Raytheon® distributor in Canada and the USA, Milltronics acquired the Raytheon Ultrasonic Ranging business segment and the AiRanger II product. Over the next 30 years, Milltronics® has become the market leader and the most trusted name in ultrasonics level measurement. After the Siemens acquisition in 2000, the Milltronics brand has combined with the Totally Integrated Automation vision of Siemens to offer ultrasonic level measurement equipment as an integral component of complete system design.

Product development map

1976   First Milltronics-designed ultrasonic measurement system, AiRanger III, installed in a cement application.

Release of MiniRanger, first compact ultrasonic system.

1978   The ST25B transducer. First transducer manufactured by Milltronics.

1981   The LR series of transducers for improved long distance measurement.

1987   The MultiRanger, the first multi-functional ultrasonic level device.

1992   The Probe, the first low-cost integral design level monitor.

1995   The Echomax series of transducers.

® Raytheon is a registered trademark of the Raytheon company.
® Milltronics is a registered trademark of Siemens Milltronics Process Instruments.

1999   The SITRANS LUC500.

2001   A new generation MultiRanger,
       the MultiRanger 100/200.

2004   The SITRANS Probe LU, a 2-wire, loop powered
       ultrasonic transmitter.

2012   The SITRANS LUT400, a high accuracy, long range
       ultrasonic controller

Ultrasonic theory

Ultrasonic measuring technology operates on the simple principle of measuring the time it takes sound to travel a distance. While the idea is simple, the process of creating, controlling, and measuring the sound travel is not.

Sound

Sound is the interpretation of electrical signals. These signals are derived from acoustic pressure waves that activate a transducer similar to the human ear. This organic transducer interprets the electrical signals channeled into the ear canal.

The sound signals are caused by the mechanical vibration of the object. The vibration is transferred to the gas modules in the surrounding medium within which it is contained. The transfer occurs as the vibrations alternately compress and decompress the molecules next to the object, spreading outward like the rings in a pond into which a stone has been thrown. As the object moves into the gas, its molecules compress into a smaller space.

As the object moves out of the gas, its molecules decompress into a larger space. This pattern or wave of compression and decompression travels outward from the vibrating object through the gas and manifests the phenomenon called "sound." If there is no gas, as in a perfect vacuum, then there will be no propagation of sound.

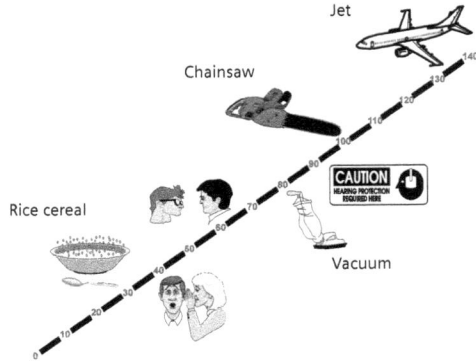

Sound levels in the everyday world

The sound, or noise, of everyday life surrounds us from our breakfast to household chores, work, and travel. Sound is everywhere and its occurrence seems a natural part of our environment. Sound, however, can also be used, not just for direct communication as in speech or music, but also as a resource to be harnessed and then applied to a method of measurement.

Using sound

Sound can be used as a measurement tool because there is a measurable time lapse between sound generation and the "hearing" of the sound. This time lapse is then converted into usable information. Ultrasonic sensing equipment has the ability to generate a sound and then the capacity to interpret the time lapse of the returned echo. It uses a transducer to create the sound and sense the echo, and then a processor to interpret the sound and convert it into information.

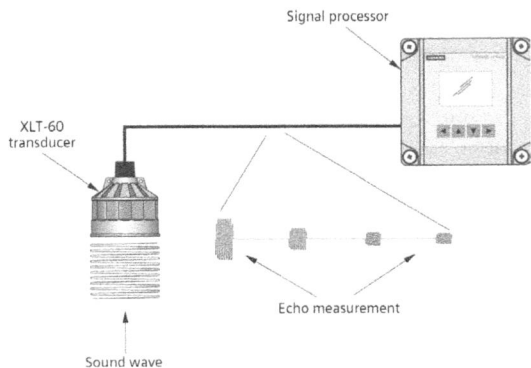

Frequency and wavelength

Vibration of the sound waves is related to time and is called "frequency." Frequency is measured in Hertz (Hz) and refers to the number of cycles per second. A pure sound wave of a particular frequency exerts sound pressure which varies sinusoidally with time. One wavelength or cycle is defined as the distance from one compression peak to the next. The wave length of a specific frequency is related to the velocity at which the sound travels:

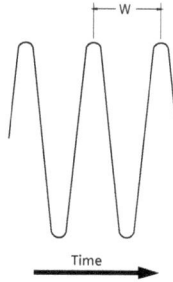

$$\text{Wavelength} = \frac{\text{Velocity}}{\text{Frequency}}$$

The number of cycles that occur in one second defines the frequency in Hertz at which the sound is being generated. For our purpose, the frequency is constant. At best, the human ear can detect sounds ranging from 20 to 20,000 Hertz. The sound range above this frequency is known as *ultrasonics*.

Measurement principle

A piezoelectric crystal inside the transducer converts an electrical signal into sound energy, firing a burst of sound into the air where it travels to the target, after which it is reflected back to the transducer. The transducer then acts as a receiving device and converts the sonic energy back into an electrical signal. An electronic signal processor analyzes the return echo and calculates the distance between the transducer and the target. The time lapse between firing the sound burst and receiving the return echo is directly proportional to the distance between the transducer and the material in the vessel. This very basic principle lies at the measurement heart of the technology and is illustrated in this equation:

$$\text{Distance} = \frac{\text{Velocity of Sound} \times \text{Time}}{2}$$

The speed of sound through air is a constant: 344 meters per second within an ambient air temperature of 20 °C. Therefore, if it takes 58.2 milliseconds for the echo to be detected, we have this result:

$$\frac{344 \text{ m/sec} \times 0.0582 \text{ sec}}{2} = \frac{20.0}{2} = 10 \text{ m}$$

The medium and the message

For an ultrasonic measuring system to have any value, it must provide a consistent output value for the same physical level conditions over a long period of time. This repeatability depends mostly on conditions of the sound media and the target material. The velocity of sound (344 m/sec) is determined through the standard medium of air and at the ideal temperature of 20 °C. However, often the conditions under which ultrasonic measurement occur are not ideal as there can be numerous factors influencing the medium, thereby altering the sound transmission speed and affecting measurement:

- temperature
- medium type (gas)
- medium stratification
- vacuum

Sound intensity

Sound intensity describes how much energy there is in a wave of sound. The units of sound intensity are watts per square meter (W/m^2). When sound intensities are compared to one another, it is usual to use the decibel as a unit of measure. The ratio of two sound intensities I_1, and I_2 is given by this equation:

$$\text{ratio in dB} = 10 \log 10 (I_1 \div I_2)$$

For sound in air, the usual reference intensity chosen as the 0 dB point is 0 dB = 10-12 W/m^2. Using that reference point, 120 dB describes a sound intensity that is 120 dB larger than the 0 dB reference intensity, which is an intensity of 1 W/m^2. 120 dB is considered the threshold of pain for the human ear. The decibel scale is used because of its ability to easily compare sound intensities which may vary over an enormous range of values.

Sound velocity and temperature

Temperature changes affect the velocity of sound in air, and the variations in temperature require compensation to calculate accurate measurement. If the temperature of the air between the transducer and the target is uniform, then compensation is achieved and an accurate measurement can be made.

The temperature of the application, or the medium through which the sound travels, is required to calculate the velocity. However, Siemens Milltronics transducers have built-in temperature sensors, and a temperature reading is taken each time the transducer is fired to compensate for temperature fluctuations.

Speed of Sound versus Temperature in Air

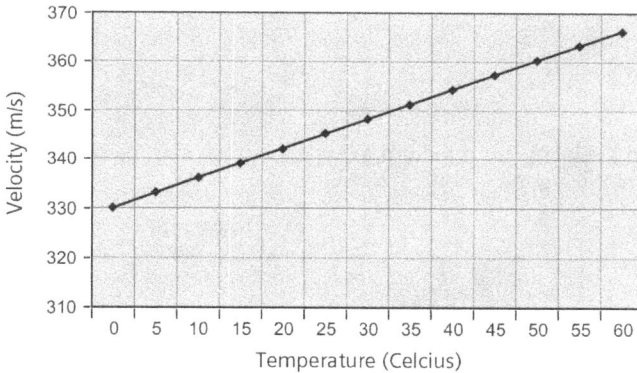

This chart tracks the increase in the velocity of sound as the temperature increases.

Sound velocity and gas

The velocity at which sound propagates in a gas is constant, as long as there are no changes in the gas. The following formula calculates the velocity for a gas:

$$V = \sqrt{\gamma RT}$$

LEGEND

V is velocity in m/sec
γ is the adiabatic index (the ratio of specific heats, 1.4 for air)
R is the the gas constant (287 J/kgK for air)
T is the absolute temperature in degrees Kelvin

9

Example

At (20 Celsius or 293 Kelvin), the velocity of sound is:

$$V = \sqrt{1.4 \times 287 \times 293}$$

$$V = 343.11 \text{ m/s}$$

VALUES

γ = 1.4 for air
R = 287 J/kgK for air
T = 293 Kelvin (20 Celsius)

Note that the speed of sound varies with absolute temperature. In air at normal ambient temperatures, which is about 300 K, a change of 1 K or C (to 301 K) causes the speed of sound to increase:

$$\sqrt{301/300}$$

$$= \sqrt{1.00333}$$

$$= 1.001665$$

GENERAL PRINCIPLE

In all ideal gases, including air, the speed of sound increases with increasing temperature by about 0.17% per °C in the range of normal ambient.

Sound velocity and pressure

Sound velocity in a medium experiencing variable pressures is calculated using the following formula:

$$V = \sqrt{\gamma \times P/\rho}$$

LEGEND

V is velocity in m/sec
γ is the adiabatic index
(the ratio of specific heats, 1.4 for air)
P is the pressure in N/m^2
ρ is the density in kg/m^3

This formula suggests that the speed of sound varies with pressure as it does with temperature.

The vapor saturation in air of various chemicals must also be accounted for. The saturation level is relevant to the different vapor pressures of each chemical as illustrated in the next chart. Note that the curved lines are for 100% saturation and the true sound velocity is in between the applicable curve and that shown for air.

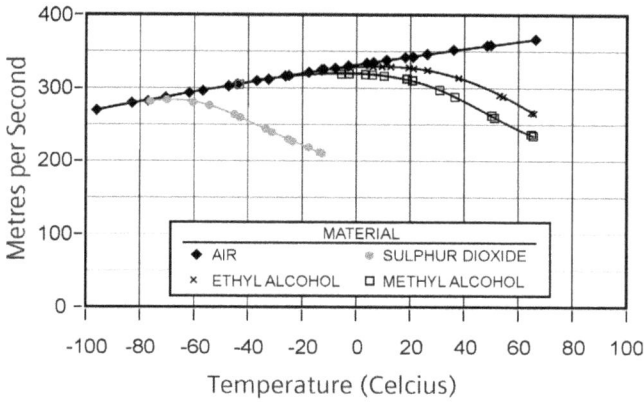

MATERIAL	
◆ AIR	◉ SULPHUR DIOXIDE
× ETHYL ALCOHOL	□ METHYL ALCOHOL

Sound velocity and vacuum

If a tree falls in a vacuum, does it make any noise? No. Sound requires something to vibrate, and in a vacuum, there is no medium to vibrate. Thus an application that operates in a vacuum has to rely on an alternate technology for level measurement.

Siemens Milltronics has a comprehensive line of radar instruments for non-contacting measurement, and a thorough range of capacitance instruments and guided wave radar for level and interface contact measurement. All these technologies operate perfectly well in a vacuum.

Sound velocity and attenuation

Attenuation refers to a decrease of signal strength as it moves from one point to another. For sound signals, a high degree of attenuation generally occurs where there are high levels of dust, humidity, or steam. Attenuation also occurs where target materials are highly absorbent to sound, foam for example. In such applications, impedance and frequency selection are essential in order to transfer as much power as possible from the transducer into the air and vice versa.

Where the medium between the transducer and the target is other than the natural composition of air, the velocity of sound can also change. If the medium is homogeneous, compensation can be achieved. If, however, the medium is stratified so the propagation of sound undergoes changes in velocity at various levels, then only an approximation can be made by using the average velocity of the medium to calculate the distance that the sound has traveled.

Sound reflection

When a sound wave arrives at an interface between media of different density (e.g. air and water), some of the sound energy is reflected and some of it is transmitted through the second medium. The ratio of energy reflected to energy transmitted is dependent upon the acoustic impedances of each media. The greater the ratio or difference, the greater the amount of energy that will be reflected.

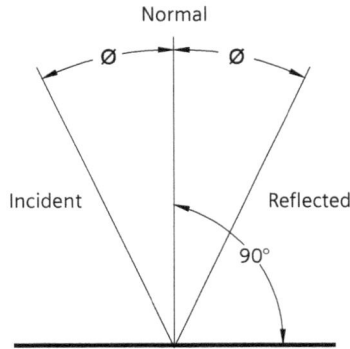

The angle of the reflected sound wave (on a smooth surface) is equal to the angle of the incident soundwave, but to the opposite side of the normal to the plane of the surface. Ideally, for measuring level, this angle is kept to a minimum.

A surface is considered smooth if the roughness, expressed as the peak to valley difference, is 1/8 or less of the incident wavelength. Any absorption of the sonic energy is ignored for this example.

Sound diffraction

Diffraction occurs when the sound wave bends around an object such that there is little or no reflection. For a given size object, diffraction decreases with a decrease in wavelength (increase in frequency).

Sound pressure level (SPL)

Sound pressure level (SPL) is the pressure of sound in comparison with the reference pressure level where P_{ref} is the reference for sound pressure in air (20.4mPa at 1KHz). The SPL can be measured by a microphone.

$$SPL = 20 \log \frac{P}{P_{ref}}$$

Sound intensity changes

When sound propagates within a gas, it spreads out so that the energy it carries is diffused over an increasing area as the wave travels further from its source. Excluding losses caused by other factors described later, sound intensity decreases at a rate that is inversely proportional to the square of the change in distance.

$$\Delta I = \frac{I_0}{(d_1 - d_2)^2}$$

LEGEND
$\Delta I = I_1 - I_2$ change in intensity
d_1 reference distance
d_2 new distance

$I_0 \qquad I_1 \qquad I_2$
$\longmapsto d_1 \longmapsto$
$\longmapsto \qquad d_2 \longmapsto$

That is to say, if the intensity of sound is X at a point I from the source, then the intensity will be X/4 at a distance of 2I from the source.

Summary

The sound waves are affected by many factors within the application environment, and the application engineer must always verify that all these conditions are known before setting up the application:

- temperature
- medium absorbency (dust, steam)
- medium type
- pressure
- medium stratification
- vacuum
- reflectivity of material

Siemens Milltronics ultrasonic instrumentation tackles applications that involve one or more of these conditions. Our experienced sales application engineers will design an instrument configuration that will provide reliable and accurate measurement.

Chapter Two

Ultrasonic instrumentation

Stop, children, what's that sound[1]

Measurement repeatability is dependent on the signal processor being used. The specified accuracy values take into account such factors as loss of resolution, supply voltage variation, operating temperature, circuit linearity, and load resistance. These factors depend on the instrumentation hardware and software, not the application conditions.

Ultrasonic level measurement instrumentation requires two components, one to generate the sound and receive the echo (transducer), and one to interpret the data, derive a measurement, and affect a reaction of the controller. Even though some ultrasonic instruments combine the components in one unit (SITRANS Probe LU, Pointek ULS200), the individual functionality remains distinct. The operation and technical specifications regarding instrument performance will be discussed in detail in subsequent chapters.

The transducer

Advances in the design of ultrasonic transducers have significantly contributed to the success of ultrasonics as a level measurement technology. Transducers are the vocal chords and ears of an ultrasonic level measurement system. The sound pulse is created by the transducer which converts the electrical transmit pulse into sonic energy, effectively radiating that sonic energy into the air and towards a target.

After the transmission process is complete, the transducer then acts as the receiving device for the returning echo signal. This information is then processed and turned into a measurement value.

The effective acoustic energy is generated from the face of the transducer and is radiated outward, decreasing in amplitude at a

---

[1] Buffalo Springfield, "For What It's Worth." *Buffalo Springfield*, 1967.

rate inversely proportional to the square of the distance as the unit energy is dissipated over a larger area. Maximum power is radiated axially (perpendicular) to the face in a line referred to as the "axis of transmission." Where off-axis power is reduced by half (-3 dB) with respect to an on-axis point equidistant from the transducer, a conical boundary is established. The diametrical measurement of the cone in degrees defines the half-power beam angle. Although the beam angle for a round face transducer can be derived empirically, it can be predicted by the following formula:

$$\sin \varnothing_h = 0.509 \left(\frac{\text{wavelength}}{\text{face diameter}} \right)$$

$$\sin \varnothing_h = \tfrac{1}{2} \text{ beam angle}$$

Transducer environments

Transducers carry a full range of hazardous application approvals from CSA and FM to ATEX (European Union Explosive Atmospheres protection). Constructed from the most advanced material compounds, transducers are available for some of the harshest industrial environments:

- For corrosive applications, transducers are fabricated with materials such as PVDF or PTFE, allowing ultrasonics to be used with acids and solvents.
- In dusty applications, acoustic impedance matching materials such as polyurethane and polyethylene foam are used because their elastic properties amplify the crystal's vibration.
- For long-range solids applications, long-range transducers deliver high power output to measure solid materials accurately to distances over 200 feet. The flexural mode transducer delivers more power by driving a large central disc with the central piezoelectric crystal. The large metal disc is made to vibrate along with the piezoelectric crystal, producing a standing wave on its surface. Holes punched in concentric rings allow every other antinode to be delayed to the point that they become in phase with the others. The net effect is an intense sound pressure wave which is transmitted into the air. This type of transducer is very well suited for dusty environments.

Transducer accuracy

The accuracy of any installation is dependent on the care taken to ensure the electronics agree with the physical measurement and the accuracy of this calibration. Due to the design of the electronics, insitu calibration is easy and high accuracy is readily obtainable. Traceability to known standards is dependent on the method and equipment used as the reference.

Transducer resolution and accuracy

The minimum change or increment of distance that can be detected is referred to as the resolution of the measurement system. Resolution is dependent on the wavelength and the timing resolution of the electronics. The shorter the wavelength, the smaller the increment that can be resolved given a specific signal processor. The SITRANS LUT400, has a design resolution of less than one millimeter (0.078") and a one millimeter accuracy specification.

In science, engineering, industry and statistics, accuracy is defined as how close the measurement system quantity is to the measurement of that quantity's actual value.

Impedance matching

The vibration of the transducer face acts upon the surrounding air to produce a sound wave. However, for efficient transfer of power from the crystal to the air, impedance matching materials must be used. Matching material steps down the high impedance of the crystal to the low impedance of air. On Siemens Milltronics transducers, a special low density material is used as an interface.

The impedance matching can be further enhanced by an additional facing material. However, this is not always required nor practical from an application standpoint.

Acoustic impedance matching is improved by these materials as their elastic properties amplify magnitude (D) of the crystal's vibration. As expressed by this formula:

$$W = F \times D$$

where W = work, F = force, D = magnitude

More amplitude is now possible with a given force by increasing the distance the vibration has traveled through the transducer face.

Axis of transmission

Sound energy is generated from the face of the transducer and radiates outward.

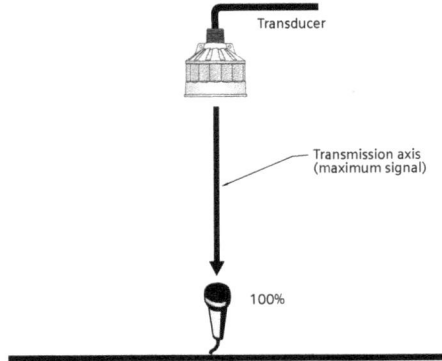

The energy generated at the face of the transducer decreases in amplitude at a rate inversely proportional to the square of the distance traveled. Maximum power is radiated axially (perpendicular) to the face in a line referred to as the "axis of transmission."

Beam width

Beam width is defined as "twice the angle at which off-axis transmission is 3 dB less than the transmission axis acoustic pressure levels (as measured equidistant from the transducer face)." Therefore, a diametrical power measurement of the cone in degrees defines the half-power beam angle.

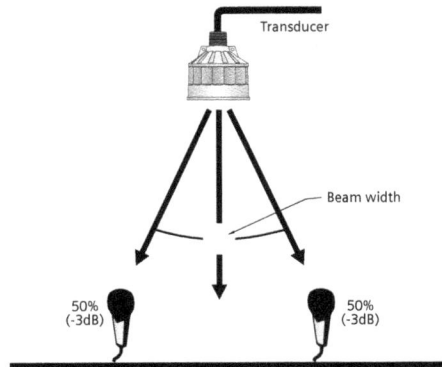

Beam width is a function of the transducer radiating surface area, frequency, and plane.

For ultrasonic level measurement, wide dispersion is undesirable. The narrower the beam width, the less likely vessel obstructions will be detected.

For short and wide vessels, a 12° beam width is ideal to simplify aiming. For tall, narrow vessels, a 5° to 6° beam width will avoid vessel wall seam or corrugation detection for maximum reliability.

Beam spreading

As well as the main beam, side lobes of a much lower intensity may radiate in the form of a conical shell, concentric to the main component. The main component and the side lobes may be depicted on a polar plot in order to visualize the pattern of sound. It is desirable to have as much energy as possible concentrated in the main beam in order to reduce unwanted echoes generated from the side lobes. Similarly, it is necessary that energy be prevented from radiating from the end opposite the transducer face. As well as good output power, the transducer must be sensitive to the weak return echoes as no amount of electronics can compensate for non-detection of an echo. Thus, proper transducer design is fundamentally important in putting the theory of ultrasonic echoranging into practice.

Ringdown

The primary active component of the transducer is a piezoelectric crystal that exhibits an expansion and contraction of its length when subjected to alternating voltage. When the voltage is removed, the crystal is no longer excited and its mechanical vibration begins to decay. The inherent nature of the crystal and the surrounding transducer mass is to continue vibrating. This vibration is called "ringing." The time it takes for this ring to stop is often called "ringdown."

The level of ringing depends not only on the crystal itself but also on the materials and construction of the entire transducer. Modern transducers have significantly less ringdown than earlier versions. Due to research into the latest construction materials and techniques, the blanking distance of the newest ultrasonic instruments like the SITRANS Probe LU is now only 0.25 meters (10″).

The controllers

The transducers can be likened to the scouts of a level measurement system. They go out and get the information and bring it back. The controllers analyze that information and then turn it into something useful.

Since its inception, ultrasonic level measurement technology has improved greatly with advances in electronic signal processors. Digital systems transmit in the same manner as analog systems, but digitize the analog received signal of the return echo and store the complete echo as a profile.

The processor, inside the controller, then analyzes the profile using software algorithms, extracting one echo from the profile as the most probable to be the target echo. Then, the signal processor converts the time differential between the transmit and the time of the selected target echo into distance.

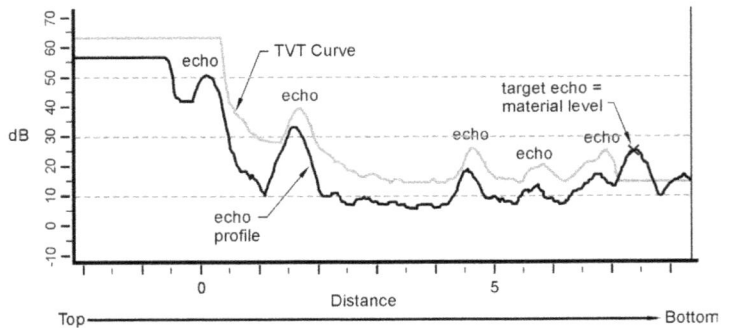

Digital ultrasonic measurement also has the ability to use software filtering techniques and intelligent echo extraction algorithms to determine distance. The microprocessor gives the signal processor the ability to perform high speed manipulation of the data gathered from an echo profile. Using analog-to-digital conversion, echo profiles received by the transducer are digitized by the receiving device and stored in memory for future evaluation. Storing the echo profile in memory makes it possible to perform the many tests on the data necessary to determine the true material echo.

Digital filtering

Digital filtering removes unwanted noise from the echo profile, including electrical noise always present in an industrial environment.

For example, variable speed motor drives produce high levels of electrical noise that are usually very high in amplitude yet very short in duration when compared to the data being gathered for the echo profile. Therefore, digital filtering is used to remove any data from the profile below a given limit in duration, significantly reducing the effect of the noise on the overall measurement quality.

Averaging echoes

In applications that create high levels of dust or acoustic noise, high speed data manipulation permits the averaging of many echo profiles to develop a composite that can be more accurately analyzed. Averaging of echo profiles performs several useful tasks: random sources of interference such as acoustic noise or air currents are averaged out of the echo data and echoes are enhanced in dusty and otherwise challenging applications.

Echo extraction algorithms

Echo extraction algorithms are software-based functions used to evaluate an echo profile. An echo profile can be evaluated in many ways, with each method having particular advantages in different applications. Before microprocessors, analog signal processors evaluated the echo profile as it returned and then the receiver looked for any echo above a set threshold. Once an echo of sufficient amplitude was detected, the distance was calculated and the output was generated. Unfortunately, analog signal processors were unable to differentiate between real but erroneous returns and the true echo in difficult applications.

Digital signal processors apply echo extraction algorithms after the entire echo profile has been received, and then use many techniques to determine which echo represents the true material level.

- One method of echo extraction involves storing a profile of the empty bin. This stored profile is then used as a template, allowing the processor to ignore obstructions in the bin. For example, in a bin fabricated with bracing around the inside surface, the return echoes can indicate bracing and not level.

  When the echo profile for the empty bin is captured, the profile shows echo returns indicating the bracings. In order to discriminate against these erroneous echoes, the signal processor compares each echo profile to the template profile initially taken and stored in memory. This template profile contains the echoes produced by the bracing. Thus, using this comparison, the echoes from the bracing are ignored and the true material echo is selected.

- Another method evaluates the echo profile based on the characteristics of the echo and its location in the echo profile. This method first selects the most likely echo in the profile by using a threshold similar to that used by the analog signal processor. Once the most likely candidates have been selected, the algorithims begin to evaluate each echo based on the type of material being measured. If the material is liquid, then the program evaluates the echo on amplitude and its location in the echo profile.

  For example, when measuring a liquid surface the characteristic echo is narrow and high in amplitude. Liquid is very reflective to ultrasonic frequencies; therefore, the true liquid level will usually be the first echo received; and in a liquid application, the algorithm would select the first echo received with the highest amplitude.

- For solids measurement, the processor selects the most likely echoes in the same manner, only using differed selection criteria based on the differing characteristics of that solid material. In this case, the program looks for an echo which is of lesser amplitude and wider than that of a liquid echo. The echo is wider because most solid materials have an angle of repose which reflects many different echoes from differing points on the angle of repose. The algorithm must now look for the selected echo which is the highest in amplitude and the widest. Therefore, the processor will select the echo with the greatest area.

Summary

Digital signal processing and advanced echo extraction algorithms make ultrasonics a reliable and accurate method of measuring solids and liquids. Ultrasonic instrumentation is thus a valuable addition to many operations, providing long term and cost-effective measurement. This book provides a thorough look at ultrasonic level technology, at the instrumentation, and at the wide variety of applications best suited for its use as this proven technology continues to be a preferred solution to many measurement needs.

Notes

Chapter Three

The sound and the slurry

Somebody's shouting / Up at a mountain
Only my own words return[1]

The transducer is the speaker and microphone in the ultrasonic level system, producing the ultrasonic waves and then sensing the echoes as they return so the controller can respond as programmed. Siemens Milltronics transducers have a proven and extensive application history, and are the reliable eyes and ears of thousands of applications around the world.

This chapter examines the role of the transducer in an ultrasonic sensing system, how it is made, and how it works.

Topics

- ultrasonic systems
- transducers
- blanking
- differential amplifiers

- echo processing
- reducing noise
- beam angle
- profiles

Transducers and ultrasonic systems

The ultrasonic sensing system is available in two formats:

- single systems: the transducer and the controller electronics are integrated into one enclosure
- compound systems: the transducer and the controller are separate entities

Single systems

A single unit system is often referred to as a "level transmitter." The Siemens SITRANS Probe LU level transmitter combines the electronics and transducer in a compact system ideally suited for liquid level measurement up to 12 meters (40 ft).

The transducer portion of the SITRANS Probe LU is in the lower half of the device; the controller,

---

[1] Deep Purple, "Pictures of Home." *Machine Head,* 1972.

electronics and wiring area are in the upper half. Level transmitters are versatile and are suitable for many applications, including both general purpose use in safe areas and use in hazardous areas, depending on approvals.

Compound systems

Compound systems separate the ultrasonic transducer from the controller. The transducer is mounted on the vessel while the controller is in a safe area away from the application in a control room or a field mounted electrical panel. Siemens offers a wide variety of controllers and transducers, like the SITRANS LUT400 and Echomax XRS-5, that can be matched to suit many applications.

Transducers carry many safety approvals for mounting and for use in hazardous areas, and they are designed to withstand rugged industrial environments. The transducer is connected to the controller by cable (either co-axial or twisted pair), receives the electrical transmit pulse, and then sends the return echo pulse along the same wire. The transducer and controller can be separated up to a distance of 365 meters (1200 ft).

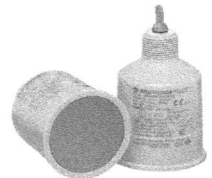

Transducers

A transducer is simply a device that converts one form of energy into another. Thus, devices such as the speakers connected to an entertainment system are transducers because they convert electrical signals generated by the amplifier into the music you hear. A microphone is the reverse of a speaker: a transducer converting sound into electrical signals.

The ultrasonic transducer performs both functions. Like a speaker, it converts the high frequency electrical pulse from the controller into high frequency sound, and then projects the sound into the vessel. And like a microphone, it converts the sound echo back into an electrical pulse, transmitting this signal back to the controller for processing. The transducer does not transmit and receive simultaneously, but constantly changes from transmit mode to receive mode many times per second.

The time required to change from transmit to receive mode is finite. See "Blanking."

The heart of the ultrasonic transducer is a piezoelectric ceramic crystal that vibrates when a high voltage pulse is applied, sending out sound waves. Conversely, when the sound waves return, the vibrations cause the piezoelectric crystal to produce an electrical signal which is then sent back to the controller for interpretation. The difference between the transmitted signal and received signal, is significant, and the outgoing transmit can be several hundred volts, while the received signal is in the microvolt to millivolt range.

Because the return signal is so slight, it can be affected by any number of situational influences: medium temperatures, attenuation, and obstructions. To achieve maximum performance benefit from the ultrasonic system, all the application conditions need to be considered when designing an ultrasonic level system.

Temperature and transducer material

The temperature of the application can affect the performance of the transducer, as does its constancy, because fluctuations also affect the reading reliability. Transducers compensate for these conditions by incorporating temperature sensors and by using temperature resistant materials in their construction so that the readings are unaffected by these conditions. The temperature variation effect is generally 0.17% for every degree centigrade; so for every degree the application temperature fluctuates, the level measurement is affected by 0.17%.

Temperature sensors

When transducers were first developed, temperature variations were mediated by the use of an external sensor which transmitted data to the controller, which then compensated for fluctuations by adjusting the reading accordingly. The need for the external sensor was eliminated when an ambient air temperature sensor was incorporated into the body of the transducer. Making the sensor part of the transducer circuitry also allows the sensor to use the same wire set to transmit temperature data to the controller. Siemens added to this convenience by placing the temperature sensor in a pocket just behind the transducer face and by improving the circuitry, enhancing sensor function by accelerating the temperature processing.

Built-in temperature compensation improves the accuracy of the system and reduces installation cost.

Sound and differential amplifiers

While noise can affect the system from outside, it may also occur within the system itself. This influence is a consequence of the

27

system's electrical functionality and the cabling requirements that create or amplify noise. Siemens has developed a differential receiver interface that eliminates or greatly reduces induced noise on both the positive and negative wires of a twisted pair cable.[2]

Figure 1
The electrical pulses received by the transceiver tend to be smaller than the initial pulses output by the device.

Transmission and receipt of both electrical and ultrasonic signals

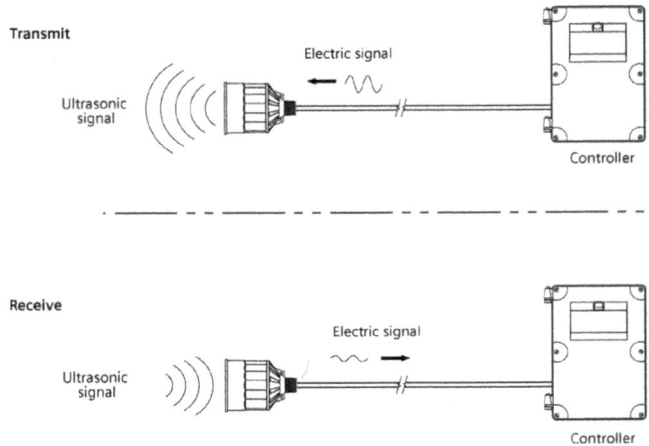

Therefore, accurate distance measurement requires amplification of the electrical signals.

For the device to calculate a distance accurately, it must amplify the returning electrical pulses and analyze the returned data using echo processing algorithms. Unfortunately, the amplification procedure used on the returning signals is sensitive to the effects of noise and this is where the differential receiver interface has a number of advantages over the common single-ended receiver interface.

Single-ended receiver

The amplifier within the controller is responsible for magnifying the returning electrical pulses existing between the amplifier's positive and negative inputs. For the common single-ended receiver interface (see Figure 2), the positive input of the amplifier is connected to the positive terminal of the transducer, and the negative input is connected to ground. When the device receives signals from the transducer, it amplifies the signal existing at the positive terminal with respect to ground. For now, if the ground is assumed to be

---

[2] This section was first published as, Aus der "Klemme" Füllstandmessung - intelligente Schaltungstechnik vergrößert Rauschabstand. Gordon Li. *MSR Magazin* (Messen, Steuern, Regeln und Automatisieren). Issue 1-2 (January/February 2005) pages: 16-17.

ideal, the output of the amplifier will simply be a magnified version of the signal returning along the positive terminal.

Common single-ended receiver connection

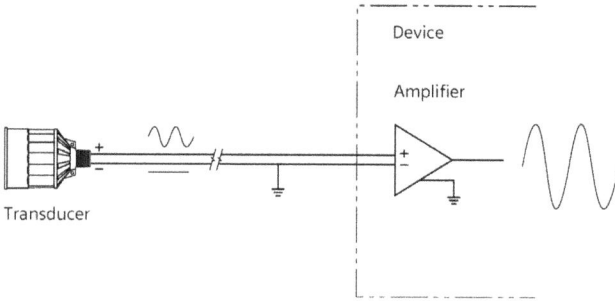

Figure 2
In a common single-ended receiver connection, the positive input of the amplifier is connected to the positive terminal of the transducer and the negative input is connected to ground.

In the event where the signal along the positive terminal is contaminated by noise (i.e. environmental noise produced by motors, near-by antennae, wireless devices, etc.), since the ground is assumed to be ideal, the amplifier would magnify this noise (see Figure 3). This noise could lead to inaccurate distance calculations by the device.

The effects of noise on a common single-ended connection

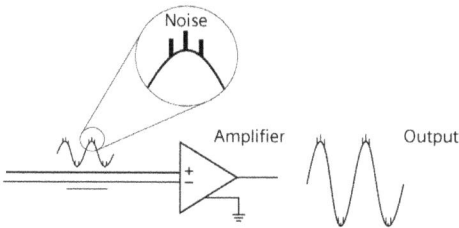

Figure 3
The effect of noise on a common single-ended connection is magnified and may lead to inaccurate distance calculations by the transceiver.

Differential receiver

In the differential receiver connection, the voltage exists between the positive and negative wires of the cable. The positive input of the amplifier connects to the positive terminal of the transducer and the negative input connects to the negative terminal.

Figure 4
In a differential receiver connection, the positive wire is connected to the positive terminal of the transducers and the negative input is connected to the negative terminal.

Differential receiver connection

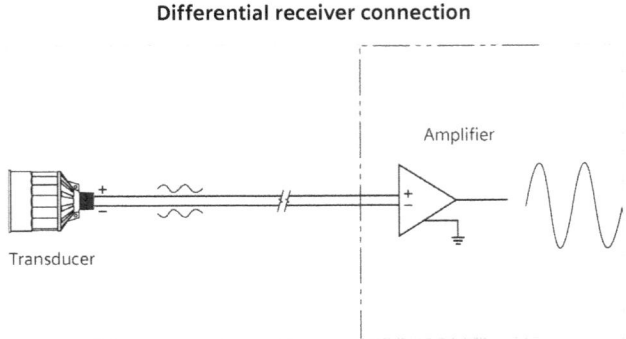

Where the positive and negative wires are in very close proximity to each other (common), any environmental noise occurring on one wire will also exist on the other. Since a differential amplifier magnifies the difference between the two wires, any noise common to both wires (hence the term "common-mode" noise) will not appear at the output of the amplifier (see Figure 5).

Figure 5
Because a differential amplifier magnifies the difference between the two wires, any noise common to both wires will not appear at the output of the amplifier.

The effects of noise on a differential connection

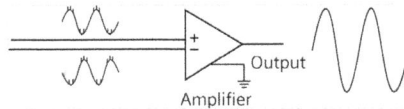

When a connection between a device with a differential receiver interface like the SITRANS LUT400 and a transducer is to be made, a shielded twisted-pair cable should be used. In this case, the positive wire connects to the positive terminal of the transducer, the negative wire connects to the negative terminal, and the shield connects to ground. Note that neither the positive nor the negative terminals are linked to ground. Since the positive and negative wires are twisted together, there is a high likelihood that the environmental noise existing on both wires will be essentially the same. Therefore, environmental noise will be present in the form of common-mode noise, which the amplifier will be able to effectively remove. Also voltages induced on the shield due to ground loops will have no effect, since the signal exists across the positive and negative wires.

Differential interface combined with the physical twisting of the wires in the twisted shielded-pair cable enhances the common-mode noise rejection ability, helping to negate noise interference.

Application temperature[3]

Ultrasonic instruments have a high temperature tolerance. For most applications, high temperature is not an issue, but in hot process applications where the material comes from a kiln or dryer, the transducer requires a high temperature tolerance. To meet these demands, design advances have extended the maximum temperature range of many transducers to 150 °C (300 °F).

Ultrasonic transducers remain extremely stable over their operating range because of their on-board sensors and two-wire data transmission, even during extreme temperature fluctuations common to many operations.

Housing material

Chemical compatibility is an important application consideration; the transducer has to be compatible with the material being measured. Transducers are available in a variety of materials, including PTFE, ETFE, PVDF, CPVC, and CFM, and can be matched with a variety of application material conditions.

Attenuation is the decrease in the sound signal as it passes through various media and the initial power/vibration of the sound is absorbed by other influences.

The user should always verify material suitability by contacting the transducer manufacturer or by using a chemical compatibility chart provided by the transducer material manufacturer. Chemical compatibility charts are also readily available on the Internet.

Range and power

The maximum range of a transducer is normally proportional to the amount of power available and the frequency of transmission. The higher the initial transmit power, the better the chance of getting an echo. The thicker the medium through which the sound travels, the more force is required to push the sound through it. Lower frequencies are less attenuated when they pass through air, which is why foghorns are so low-pitched.

Ranges quoted in the specification sheets and instruction books should be taken as a maximum. Do not exceed!

---

[3] Doug Duncan. "Ultrasonic sensors: Now an even better choice for solid material detection," *Instrumentation and Control Systems*. November 1998.

**Signal attenuation round trip
70° F, 50% Relative humidity**

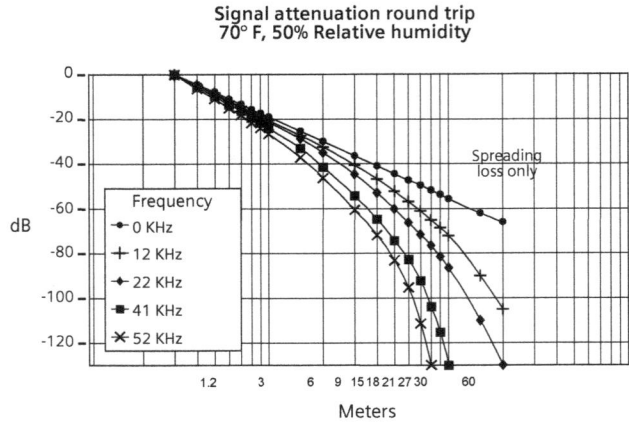

The materials present in the medium absorb (attenuate) the sound and affect performance. This graph[4] shows the attenuation in decibels of various sound frequencies when the temperature and humidity levels are constant:

So at 41 kHz, the round trip attenuation is approximately 55 dB at a range of 15 meters (50 ft). An echo of 30 dB at the maximum rated range of a transducer is strong enough to process and exceed the noise levels inherent in many applications; however, in some instances, echoes as weak as 5 dB can be processed with high confidence. Knowing this, the controller must send out a pulse to the transducer that is strong enough so that in the after-round-trip attenuation of 55 dB at 41 KHz at a 15 meter range, there is sufficient return echo strength to process the signal.

Long-range transducers are designed to operate at lower frequencies to take advantage of lower attenuation rates that provide a stronger echo to the transceiver for processing. In exchange for the extra power, however, some resolution is lost. But in long-range applications of grains, powders, or pellets, this generally does not present a problem as accuracy requirements may not be as stringent.

---

[4] From "The Theory of Ultrasonics and Echo Ranging."

Conditions

Conditions affect performance because the sound waves need to go through the medium and are influenced by other occupants of that space. Dust, steam, and high humidity attenuate sound and the required distance for measurement may not be met. Without an echo, there is nothing for the electronics to work with. Conversely, with a very small stainless tank, it may be better to go to a lower power transducer because multiple echoes may require changing the algorithm, an inconvenience to the customer.

Rule: if the application is on the limit of the transducer's range, step up to the next transducer for better reliability!

Dust

Most solid applications are dusty. In areas where dust is extremely heavy, keeping the transducer clean is critical because material buildup reduces the transducer's ability to transmit and receive. Siemens transducers use materials with good release characteristics which, combined with pulsating displacement, often prevent build-up from beginning.

Stilling wells

If an application is very agitated, dusty, or has a lot of surface foam, these conditions may be neutralized using a stilling well. The stilling well is a secondary pipe accessing the tank contents and reflecting the same content level.

In a very small stainless tank, go to a lower power transducer because multiple echoes may require changing the algorithm.

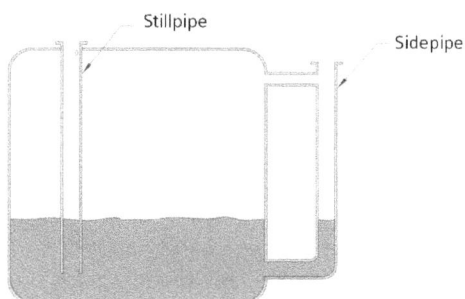

Stillpipe

Sidepipe

With stilling wells, lower power may be better since there is no spreading loss because the sonic energy is contained within the pipe. In laboratory conditions equivalent to an empty warehouse, or outside on a calm day, echoes have been obtained at much greater distances than the nominal rating of the transducers. However, in real-world applications, use the guidelines detailed above.

Foam facing

Impedance matching foam facing can improve a transducer's power and range. Impedance matching provides greater acoustic power output and is very useful in dusty situations where the particles in the air cause attenuation.

An impedance matching polypropylene closed-cell foam face is designed for Echomax XPS transducers. The closed cell polypropylene has a high temperature rating of 95 °C (203 °F) and is not bothered by dust or moisture getting into the foam.

Moisture on transducer face

The face of the XPS transducers is active and will ultrasonically evaporate the droplets. If there is a bit of a slope to the transducer (mount a degree or two off vertical), the extra vibration will cause the droplets to move to the edge and fall off.

Transducer selection

Selecting the right transducer affects performance, and the criteria should be based on range (distance) and power. For example, in some powder and plastic pellet applications, a shorter range Echomax XPS-15 (operating at 44 KHz) is more appropriate than a long-range Echomax XPS-40 (operating at 21 KHz) because the higher frequency yields a better reflection from the material surface. All the application conditions need to be considered, not just distance.

Blanking distance and height placement

Proper aiming, location, and mounting are other important factors.

The location of the transducer also influences performance. The heart of a transducer is a piezoelectric crystal; when a transducer is sent a transmit pulse, the piezoelectric crystal expands and contracts lengthwise to produce the ultrasonic sound signal. When the transmit pulse stops, the crystal is no longer excited and its mechanical vibration begins to decay. This decay has a finite time period and is often referred to as "transducer ringdown" during which the transducer cannot receive an echo.

The time of the ringing is then converted to distance (D=VT) and represents the space in front of the transducer face where measurement is not possible. This space, also known as the "blanking distance" or "dead zone," has decreased dramatically over the years as technology and research advanced ultrasonic measurement technology. Initially, the blanking distance was a meter or even more, but it has been reduced to 25 centimeters (0.82 ft) in the most recent systems.

Other factors like temperature and installation techniques also affect the blanking distance:

The ringdown quantity depends on the crystal itself as well as the materials and construction of the entire transducer.

Temperature

High temperature transducers have a longer blanking distance. While ringdown takes the same amount of time regardless of the transducer's temperature, the speed of sound increases with temperature. Therefore when using D=VT, a greater velocity results in an increased distance. The blanking distance is stated at the maximum transducer temperature rating.

Installation

If the transducer is connected directly to a metal coupler or if the mounting threads are torqued too tightly, the mounting arrangement acts as part of the transducer and will "ring" with each transmit pulse. To reduce ringing only, hand-tighten the transducer into a blind flange or onto a conduit adapter. If possible, use a plastic conduit adapter rather than a metal one.

For a flanged transducer, use the supplied plastic bolts and do not over-tighten. Use a gasket between the vessel flange and the transducer flange.

Transducer design: the heart of the matter

The physical design of today's transducer houses decades of research and enhancements, distinguishing Siemens from the rest of the players. These transducer developments have made Siemens the best in the business and the world leader in the field.

Measuring distance is not as simple as timing the sound when it bounces back. It also requires intelligence because sound is indiscriminate. It bounces from anything and everything, and since most applications are filled with obstructions and other clutter that interferes with the path of sound, it requires sophisticated equipment to distinguish the real level echo from the rest of the bouncebacks.

Siemens ultrasonic instrumentation has the sophisticated technology that evaluates the echoes and determines which ones are true and which are false.

Summary

The theoretical principle of sonic level measurement is simple. A transducer located at the top of a bin or well transmits a pulse of sound energy. The sound reflects from the material surface and travels back to the transducer where it is converted into an electrical pulse. The time delay from transmission to received echo is converted into distance and the material level is determined.

In practice, sonic level measurement is a bit more complicated as there is not just a single echo from the material, but many echoes reflecting from bin walls, beams, pipes, and wires in the sound path. Furthermore, the actual echo qualities can vary greatly for the following reasons:

- sound attenuates with distance; thus, the desired echo from the material surface may be weaker than the undesired echoes
- the material surface of solids is usually inclined, and this spreads and weakens the echo
- electrical noise from motors or controllers can interfere or be stronger than the echo

Distinguishing among the many echoes can thus be very difficult, but it is crucial for the system to be effective and the real echo needs to be found. The historical progression of the technology is based on this search for the true echo.

Chapter Four

Echo processing

Making good decisions is a crucial skill at any level.[1]

Ultrasonics technology has been used for industrial bin level measurement since the 1940s. Since then, the tools have evolved to overcome the acoustic problems faced in real world applications. Complex circuitry with time varying gain, automatic gain control, range tracking gates, near and far blanking, noise suppression filters, phase locked loops, and other advances have expanded the capabilities of sonic level measurement. These developments all have one thing in common: real-time analog processing of the signal as it was received.

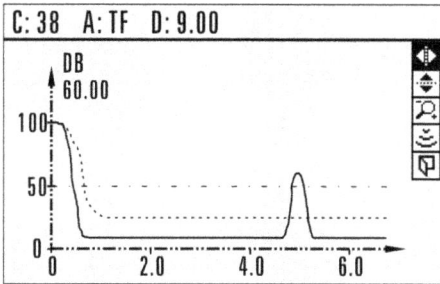

While analog processing greatly advanced ultrasonic technology as level measurement instrumentation, it also had two main disadvantages:

- First, a skilled technician with a screwdriver and an oscilloscope would carefully adjust the installed electronics for each application, and this had to be done while the bin level was cycled up and down several times. This is a costly and time-consuming process.
- Second, since the signal is processed as it is received, there is no opportunity to analyze and compare the information statistically to see if it is even possible or correct. Thus data logging was not possible.

---

[1] Drucker, Peter. 1909-2005.

Siemens recognized that analog-based instruments would never be sufficient and focused its efforts on moving away from real-time processing. Engineers soon realized that placing the entire stream of echoes into a stored echo profile in a computer's memory gave the unit the opportunity to examine its shots. The computer then processes and analyzes the stored echo profile to select the true echo in a more reliable fashion. The first product to use this revolutionary technique was the Milltronics AiRanger IV, introduced in 1984, and this system has been in every instrument since then, including SITRANS LUT400 and SITRANS Probe LU. As a result, echo-processing adjustments are not generally required in most applications.

This chapter presents Sonic Intelligence functionality and how it refines the analytic process determining true echo and level.

Topics

- Echo processing
- Shots and profiles
- Filters
- Time varying threshold (TVT)
- Echo selection

- Figure of merit
- Echo profiles
- Echo processing parameters
- Noise

Echo processing - intelligence

The patented echo-processing technology embedded in ultrasonic level measurement products from Siemens is known as Sonic Intelligence™. Sonic Intelligence differentiates between true echoes from the material being measured and false echoes generated by obstructions or electrical noise. The result is repeatable, fast, and reliable measurement.

Sonic Intelligence is field-proven in over 1,000,000 applications worldwide.

This technological advancement was developed in consultation with field service engineers and supported by field data gained from decades of experience with applications in many industries and continues to evolve today. Using higher order mathematical techniques and algorithms to provide intelligent processing of echo profiles, this "knowledge-based" technique produces superior performance and reliability.

™Sonic Intelligence is a trademark of Siemens Milltronics Process Instruments.

SITRANS LUT400 uses next generation Sonic Intelligence for echo processing, providing adaptive digital filtering of the transducer signal. For example, when noise levels are high, filters are adjusted to maximize the signal to noise ratio. This advanced Sonic Intelligence not only allows for better filtering, but provides improved tracking of echoes, and more sophisticated echo positioning algorithms.

Understanding echo processing

Echo processing is a simple concept and a complex reality as the instrument looks at each grouping of echoes (a shot) and processes the return to determine the true echo.

The short shot uses a narrow transmit pulse which reduces transducer ringing to provide improved reliability for the detection of near targets. The long shot uses a wide transmit pulse to provide a stronger return echo to detect more distant targets. Both short and long shots are processed, and then the best is selected with a bias in favor of the short shot. Once an echo is selected, various filters are applied to stabilize the reading.

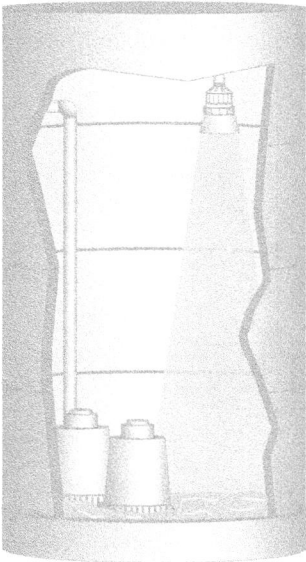

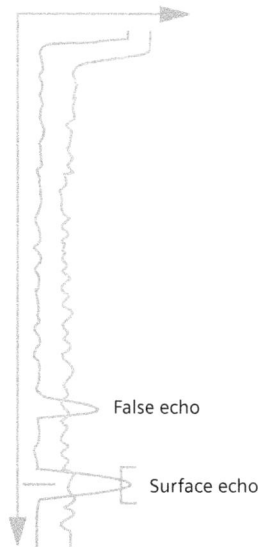

False echo

Surface echo

Even though Sonic Intelligence will select the correct echo after every measurement, the reading itself may fluctuate as a result of changing conditions within the bin. Averaging the readings reduces the fluctuations but causes a lag in response when the material level changes. Sonic Intelligence uses powerful statistical techniques, to give stable readings while still allowing rapid response to actual changes in the material level.

Shots and profiles

So let's look at how the software reads the shots and what it sees as the following are implemented:

- filters
- Time Varying Threshold (TVT) application
- echo analysis

The echo profile

The complete profile including transmit and ringdown is displayed and the amount of blanking indexed in the software is shown on the bottom trace. The echo lock window (or window) locks on to the selected echo in order to ignore spurious echoes from agitator blades or random electrical interference. A Time Varying Threshold (TVT) curve provides a reference for comparing echoes.

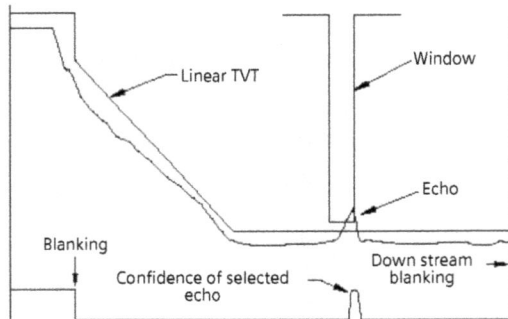

SITRANS LUT400 uses next generation Sonic Intelligence for echo processing, providing adaptive digital filtering of the transducer signal. For example, if noise levels are high, filters are adjusted to maximize the signal to noise ratio. This advanced Sonic Intelligence not only allows for better filtering, but provides improved tracking of echoes and more sophisticated echo positioning algorithms.

What the LUT400 sees (or, more accurately, "hears") from a transmit pulse in an application is stored in a digital format and is referred to as an "echo profile." This profile is a digital representation of the returned echo energy versus time.

Finding the true echo

As the echoes bounce back from the surface, Sonic Intelligence software parses the information in its goal to find the true echo. The process comprises three active components working together in the selection process:

1. Filters: enhancing the echoes and using Time Varying Threshold (TVT) to select.
2. True echo: selection.
3. Selected echo: verification.

1. Filters

Filters are algorithm processes in the software that tune out outside interferences and enhance the echoes returned from the application.

Clipping filter

Removes the high spikes in the time domain, resulting in less ringing in the filter and reducing the impact of high noise on the probability of picking the right echo.

Digital band pass filter

This method permits the design of the optimum matched filter, creating the highest probability of selecting the right echo over noise. Each transducer has a band pass filter based on its transmission frequency.

Kalman filter

This algorithm smoothes and applies damping to readings, removing jitter and instability.

Temperature filter

A Kalman filter algorithm that removes jitter of the reported temperature value contributes to accurate calculation of the sound velocity.

Spike filter

Electrical noise produces false echoes which are narrow and high compared to wider echoes from the material being measured. The spike filter in Sonic Intelligence uses this difference to identify and remove electrical noise.

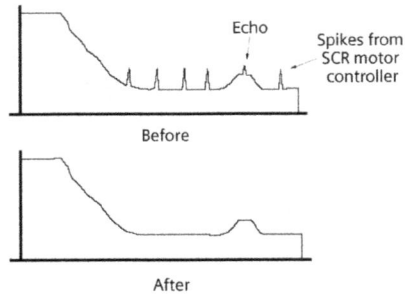

Before

After

The spike filter measures the width of each spike and cuts off any spike too narrow to be an echo. The echo lock window is the first line of defense against random electrical noise, but the spike filter effectively corrects for electrical interference from machinery such as variable frequency drives (VFDs). The spike filter will never remove a valid echo.

Narrow echo filter

Many bins have targets in the path of the sound wave that produce false echoes that could interfere with the measurement. These may be pipes, cables, or even seams in the bin wall. In a narrow bin, it may be impossible to move the transducer far enough away to avoid these targets. These targets produce echoes which make it difficult to select the correct echo from the material surface in a reliable manner.

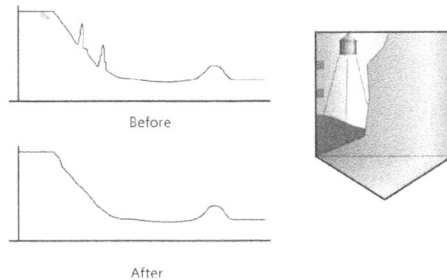

Before

After

In solids, the Narrow Echo Filter can remove undesired echoes from hard targets such as pipes, beams, windows, or seams. That's because these echoes are narrower than the true echo from an inclined solid material surface. In use, the filter is increased by small amounts until the undesired echoes are eliminated. It is advisable to monitor operation over a complete fill and empty cycle to avoid setting too high a filter value that could wipe out the true echo as well. This technique does not apply to liquid measurement where the true echo is as narrow as the undesired echo.

Reform echo

In deep bins, the sound wave faces many obstacles in its journey from transducer to material surface and back again. These obstacles include reflection from the bin wall, dust, air currents, changes in air density, and irregularities at the material surface. All these obstructions combine to create a fragmented or splintered echo. Reform echo builds the fragmented echo back into a single, strong, well-defined echo pulse.

By increasing reform in small amounts, the operator can join the peaks of a fragmented echo to produce a large, smooth echo. Use this feature with caution because if the reform is too large, the peaks of unrelated echoes will be joined together.

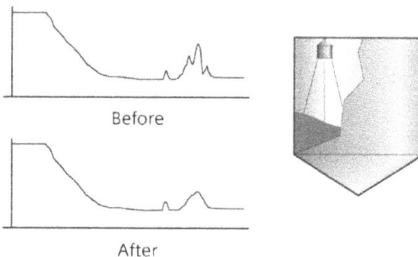

Before

After

Echo selection algorithms

After the echo profile has been enhanced, the final task is selecting the best echo, a task determined by the bin conditions.

2. True echo selection (selection of echo reflected by the intended target)

This selection occurs when the portion of the echo profile above the Time Varying Threshold meets the evaluation criteria of Sonic Intelligence. Insignificant portions of the echo profile outside of the measurement range (e.g. Far Range) and below the TVT curve are automatically disregarded. The remaining portions of the echo profile are evaluated using the echo select algorithm and the echo profile portion providing the best echo (confidence) is selected.

NOTE: A confidence value is a static test of a single snapshot profile to maintain a valid reading; it requires that each individual profile maintains its peak above the threshold. The window may be locked on the profile for hours or even days, so if the profile drops below the TVT curve just once, loss of echo may occur.

Time Varying Threshold (TVT)

The time varying threshold (TVT) is used as a reference to compare echoes at different positions in time. A far echo will be weaker because the sound has traveled a greater distance, but near and far echoes must be compared in order to select the true echo.

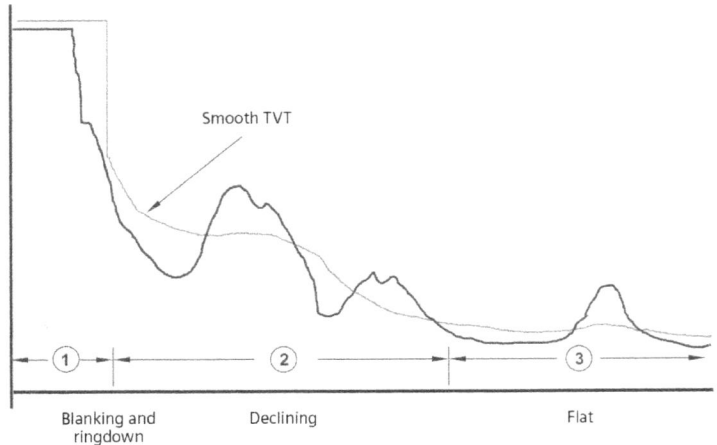

Smooth TVT

① Blanking and ringdown　② Declining　③ Flat

For a given condition of dust, temperature, or humidity, the attenuation of sound is usually uniform throughout the bin; therefore, the ideal TVT is a line with a slope equal to the attenuation of sound in dB/m. This slope is calculated from the declining portion of the echo profile.

TVT shaper (manual adjustment)

The shaper divides the TVT curve into equal segments, each of which can be adjusted up or down. An undesired echo can be ignored by lifting the TVT curve at the position of the echo. If the TVT curve is raised too much, however, the true echo will not be detected when it is in this region. It is not necessary for the TVT to clear the undesired echo, just cutting it in half is usually enough to de-emphasize the undesired echo.

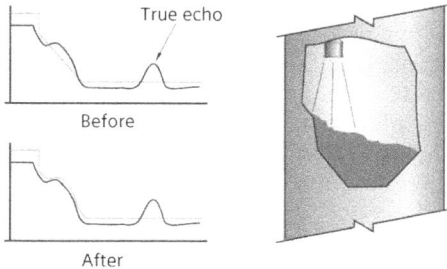

True echo

Before

After

The adjustments ride up and down with the TVT and move horizontally when the velocity changes. All calibration points and ranges should be set before using the shaper.

The shaper cannot be used on echoes from reflections of material hang-ups because the false echo position will change if the material falls out of its hung up position.

Auto false-echo suppression (automatic TVT shaping)

This feature automatically detects and suppresses echoes from vessel obstructions. Prior to activating the Auto False-Echo Suppression, empty the tank as much as possible. If there is an agitator, it should be operating. During the TVT Learn session, the controller fires pulses and shapes the TVT around any echo from an obstruction like the agitator. This shaping is easy to perform and does not require a PC.

The TVT will then not drop lower than the learned position.

SITRANS LUT400 has advanced tracking ability and can find the real echo amongst stationary clutter echoes. Therefore, even if the echo drops below the TVT curve, it can be identified with near certainty for approximately 30 seconds. This capability is measured by the Figure of Merit (FoM) setting.

There are four selection methods:

- measure the area of each echo
- measure the height of each echo
- measure the first moving echo
- select the first significant echo

For the MultiRanger, Largest or First are the preferred algorithms. For the AiRanger, together it's ALF.

These methods can be referred to as Area, Largest, and First algorithms (ALF). Another method uses Tracker where the user selects/ deselects the algorithm standard settings. Sonic Intelligence uses all four algorithms on each measurement. For each algorithm, a score or confidence is calculated for each echo that crosses the TVT curve, then the scores are added up to give a total score for each echo. The echo with the highest total score is deemed to be the true echo from the material surface. The net echo confidence is the confidence of the chosen echo minus the confidence of the echo with the next highest confidence level; i.e. net confidence of the first choice is first choice minus second choice.

Occasionally, one algorithm will give a low score to the true echo and a better score to a false echo, but the other algorithms will pull the total score of the true echo up, and push the total score of the false echo down. The final selected echo is correct because Sonic Intelligence incorporates a diversified portfolio of application experience. For a given bin condition, a single algorithm will give higher echo confidence but, as bin conditions change, another algorithm may be required. Generally, using all four algorithms provides for more reliable operation over time. However, trained operators can adjust the device parameters and fine-tune the algorithms to suit their application.

Siemens Milltronics equipment also contains an echo processing algorithm call "bLF," also known as "Best of Largest at First." This algorithm runs the Largest algorithm and First algorithm and selects the echo with the highest confidence value. It does not sum the values as in the ALF algorithm. bLF is the best choice for short range liquid applications.

Another algorithm is True First (TF). This algorithm selects the first echo that crosses the TVT curve. Use in liquids applications free of obstructions when confidence of first echo is high.

The TR (TR) Tracker algorithm selects the echo that closest to the transducer and is moving. An echo that is below the TVT line and moving can be detected by this algorithm and when it is not track-ing an echo, it behaves the same as bLF.

NOTE: Only use TR algorithm in process applications with either of these two conditions:
- continuous level changes
- a risk of fixed obstructions that could interfere with true level resulting in low confidence.

3. Selected echo verification

This verification is automatic. The position (relation in time after transmit) of the new echo is compared to that of the previously accepted echo. When the new echo is within the Echo Lock Win-dow, it is accepted, updating displays, outputs, and relays. If the new echo is outside of the window, it is not accepted until echo lock requirements are satisfied.

These three processes work away in the background as the Sonic Intelligence software determines the best echo. However, the echoes can be further tuned should it be necessary.

Echo quality

Echo quality and confidence can be further tuned using the infor-mation provided by the device from the following features.

Figure of merit

Figure of Merit (FOM) measures the quality of the reported process value: higher values represent better quality. Even when a low con-fidence value exists, a high FOM will ensure the true echo has been selected. Approximately 20 readings are used to support the FOM value.

> *EXAMPLE* FOM greater than 75% = good quality
> FOM less than 50% = poor quality

47

Contributions to the FoM:

- success of the tracking (how closely can the next level versus the actual next level be predicted)
- level of noise
- confidence of the last echo time interval since last valid echo
- speed at which the process is moving
- quality of the echo shape and how it helps the calculation of the echo position

NOTE: If FoM is low, reduce the noise in the process, or check the installation to increase signal quality.

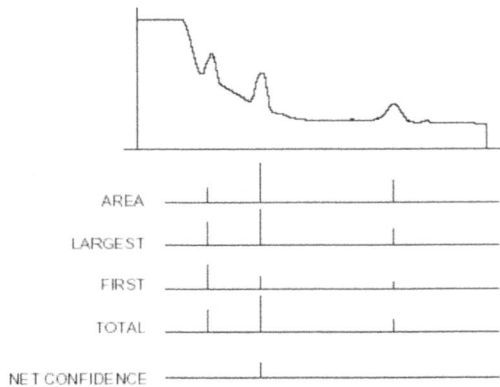

Echo parameter fine tuning

Sonic Intelligence software eliminates the need for echo-processing adjustments in all but the most difficult applications (less than 1%). If fine tuning is required, operators can use the more advanced adjustment parameters available in Siemens transceivers.

Only one or two echo-tuning parameters should be adjusted at a time, so that the impact of each adjustment can be carefully monitored for a complete fill and empty cycle. Adjustments made to reach a confidence figure at a certain bin level may not hold under changing bin levels. Do not rely solely on the confidence figure, but look at the echo profile and decide if the echo is well defined under various conditions and will provide reliable tracking of levels from empty to full. This is especially important during aiming and mounting.

Echo profiles

The echo profile, a graphic representation of the reflected echo received by the transducer, is an important diagnostic tool. Digitally generated, the profile is stored by the transceiver and can be viewed using several types of configuration software, including SIMATIC PDM (Process Device Manager), AMS, FDTs, Dolphin Plus, and Super Sonex.

The echo starts out as an analog signal:

Normal transmit pulse   Echo

This signal is digitized by the transceiver's A/D converter and changed into an echo profile, which is a measurement of the echo energy versus time.

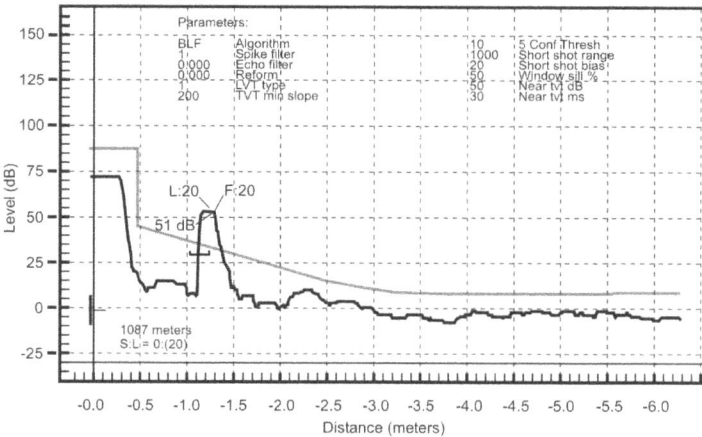

To read the profile, start at the left side, which represents the tank top, and move across. The farther right, the farther down into the tank the shot goes, reading the distance. The upward spikes indicate echoes and the strength in decibels.

Profile components

There are several components to an echo profile, each revealing a significant aspect of the echo. The echo profile components are shown in the following diagrams:

49

Echo profile

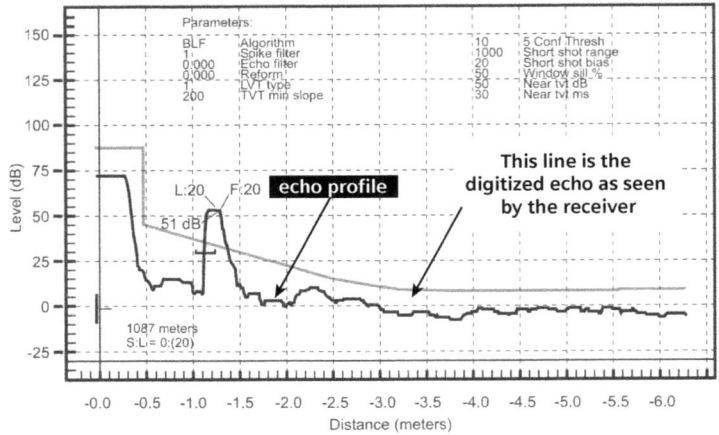

Ringdown

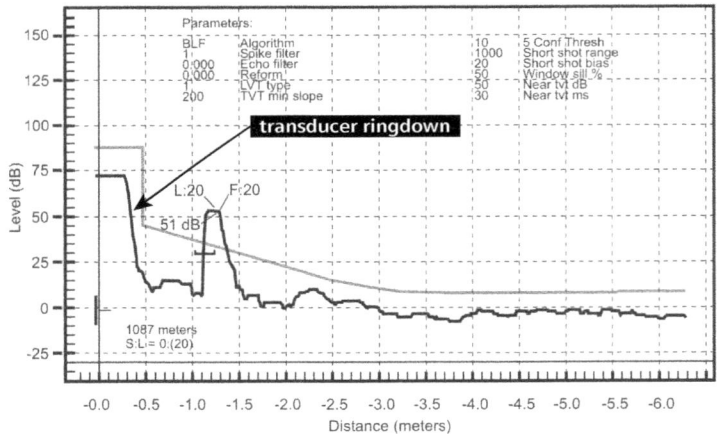

The leading edge of the echo profile is the decay of the transmit pulse, also known as the "transducer ringdown." This is the portion of the echo profile showing the gradual recovery of the crystal after a shot has been taken.

TVT curve (Time Varying Threshold)

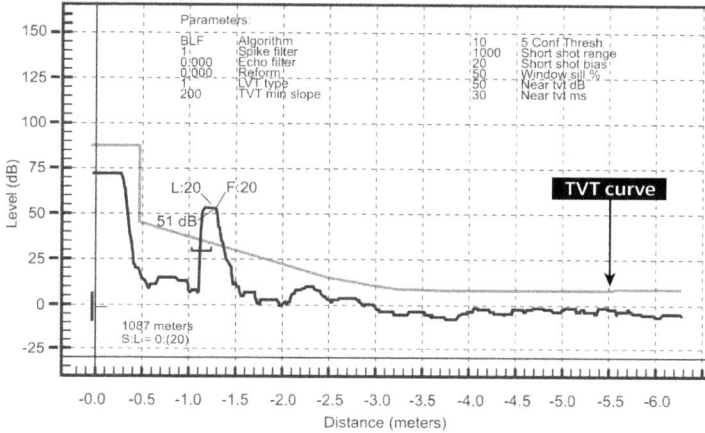

The TVT curve is a digitally generated threshold level that acts as a validation point for the received echoes. It is a dynamic threshold level; any part of the profile above the curve will be included in the echo validation process.

Echo marker

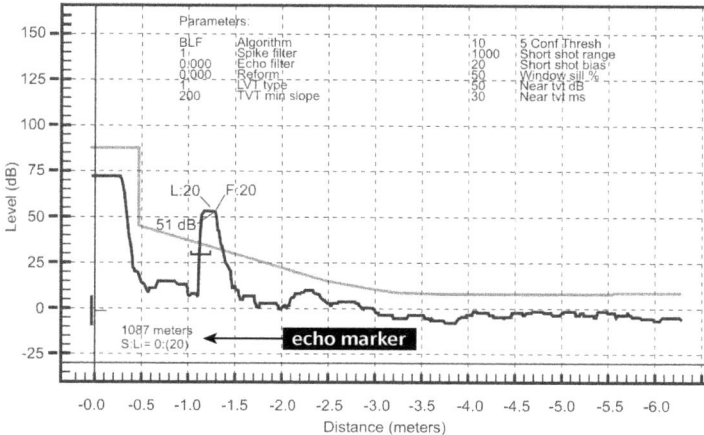

The echo marker represents what the controller has processed as the correct echo (or material level) for that particular echo profile. It may or may not present this as the level reading. The Echo Lock

Window controls the level reading output. The height of the marker represents the echo confidence; the larger the height, the better the confidence that this is the echo that most represents the true level for this particular profile.

Echo lock window

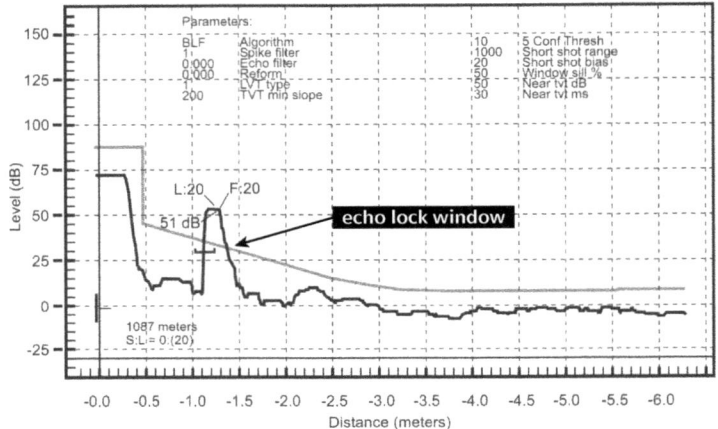

The Echo Lock Window (or simply the Window) controls how far the level can change between shots. From shot to shot, the level change cannot exceed the width of the echo lock window. If the echo marker falls within the width of the window, the controller will output the level as indicated by the echo marker and re-center the window on the echo. If the echo marker falls outside the window (e.g. marks an echo from a blade of an agitator), then the controller will ignore the marked echo and output a level represented by the window (which also represents the last valid reading).

The width of the window is set by the measurement response (or damping) parameter. If the application level changes rapidly (a process vessel), program a fast measurement response and the window width will be quite wide. If the application has a slow moving level (a storage vessel), then set the measurement response parameter to slow and the window width will be quite narrow.

If there is a sudden change in level and the echo marker (this time representing the true level and not a false echo from an agitator blade) is outside the width of the window, the controller maintains the level output at the level represented by the window, the last valid reading. It also begins to widen the window at the measurement response rate. If the echo marker points at the same echo outside the window for five shots in a row, the controller accepts that this is now true level and changes the level output (the display, mA output, or digital output) at the measurement response rate in the direction of the true level. The window continues to widen at the measurement response rate and once the true echo is within the window width, the window will shrink back to its normal size and re-center itself on the true echo.

Echo processing parameters

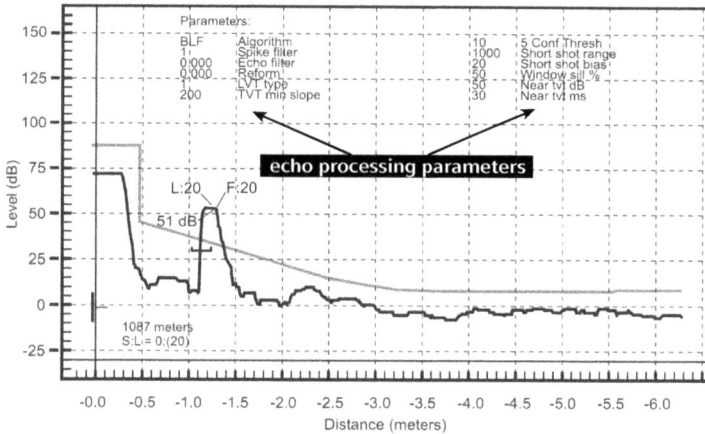

The values of some of the advanced echo processing parameters are placed on the echo profile as a convenience and can be used as a reference when subsequently viewing the profiles.

Included are the values for the following:

- echo algorithm
- spike filter
- narrow echo filter
- echo reform
- TVT type

- confidence threshold
- short shot range
- short shot bias
- window sill percentage
- near TVT dB

Echo confidence

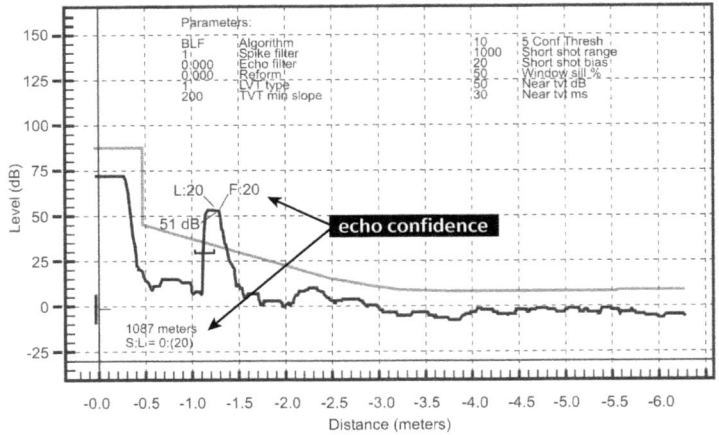

Echo confidence represents the certainty, or confidence, that the chosen echo is the correct echo. The value is generated from one of the echo processing algorithms like ALF or bLF *(see Sonic Intelligence)*. Confidence is shown in two areas: at the selected echo and beside the echo marker. The values beside the echo show the confidence values generated by individual algorithms. In the above profile it shows L:20 and F:20 – meaning that the Largest algorithm and the First algorithm each generated a confidence value of 20. The confidence value beside the echo marker is shown as:

$$S:L = 0:(20)$$

This indicates the short shot confidence is 0, the net long shot confidence is 20, and the brackets around the 20 indicate that the long shot was used to process and output the level measurement reading.

The echo

Liquid

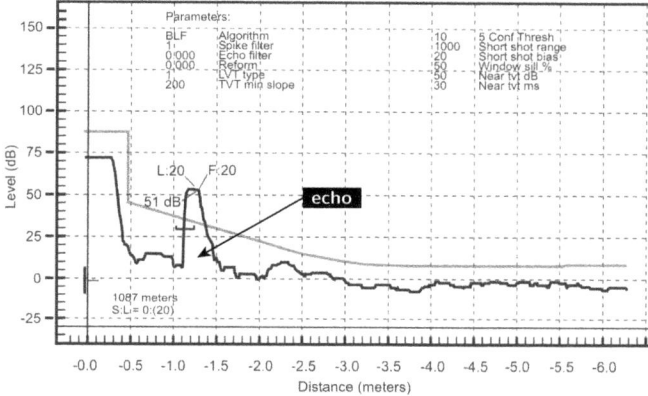

This graph shows an echo, with spikes that are sharp or gradual (depending on the material) rising in the echo profile. Sharp echoes arise from flat material surfaces like liquids, as in the above profile. Gradual, wider echoes come from solid material; clinker, coal, or grains, are more gradual in their rise, and often look like the echo below from plastic pellets:

Solid

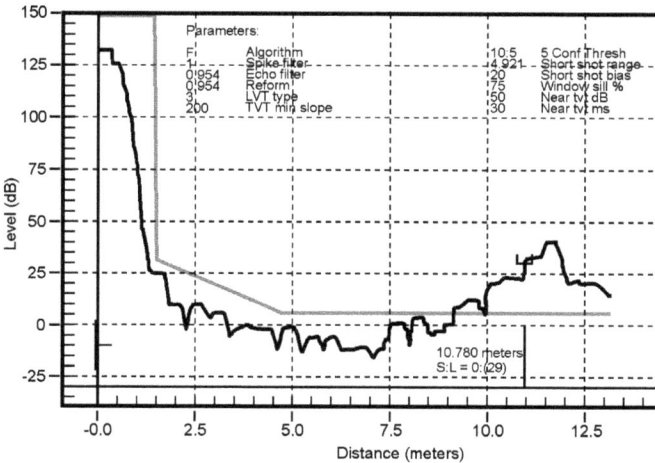

Echo strength

The strength of the selected echo in dB above 1 microvolt RMS is calculated as echo strength.

The echo profiles are an invaluable asset when troubleshooting, and a technician can determine the following from the profile based on the echoes:

- good echo strength and low confidence means multiple echoes to get the level reading
- low echo strength and low confidence can mean the echo is weak, the result of poor aiming, or high attenuation from dust, steam, or CO_2. The surface may also be foamy or poorly reflective from extreme turbulence.

Noise

Noise in an ultrasonic system can also be generated by electrical interference received from the transducer cables. This interferes with the echo reading and enters the controller from many sources:

- main power
- mA output
- external temperature sensor (if used)
- radiated from nearby electrical devices

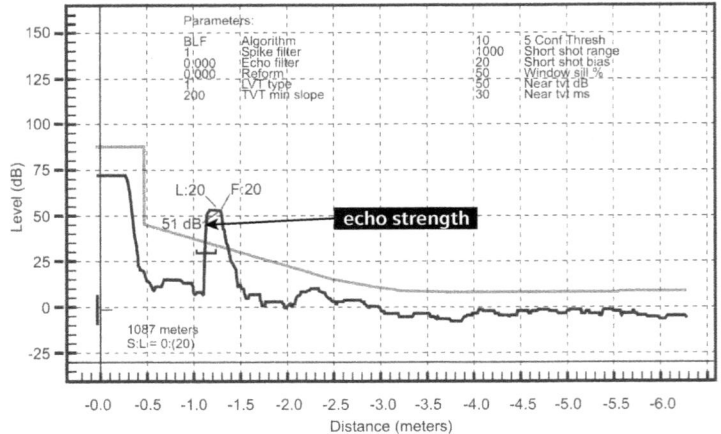

The electrical noise is picked up on the profile. The two lines at the bottom left represent the noise parameter. The top of the bold vertical line is the peak noise in dB; the small horizontal line indicates the average noise. In this example, the peak noise is 5 dB and the average noise is - 3 dB.

The best method for ensuring the lowest noise possible is to use the recommended transducer cable. The newer Siemens Milltronics controllers (SITRANS LUT400, MultiRanger 100/200, HydroRanger 200) use shielded twisted pair cable while established models (SITRANS LU) use RG62 A/U coaxial cable. The shields must be properly connected, and use grounded metal conduit to shield the cable from electrical noise and to physically protect the cable from damage.

In general, average noise levels above 30 dB are cause for concern and noise should be reduced.

The levels of noise present at the receiver input can be viewed on the device Local User Interface as the average noise and the peak noise. In general, the most useful value is the average noise because the peak noise is a spike that can be filtered out. Average noise, however, is the overall baseline noise. If the echo strength does not exceed the baseline noise, there will be problems discerning the echo from the ambient noise.

Adaptive profile blending

When not in a high noise application, the LUT400 processes every shot individually. If noise levels increase above a threshold of 35 dB, the SITRANS LUT400 blends multiple profiles together, increasing the reliability of echo selection and position detection. This replaces the previous process of averaging a number of shots and responds more quickly to fast changing material levels.

Noise interference

With no transducer attached, the noise is under 5 dB, which is called the noise floor. If the value with a transducer attached is greater than 5 dB, then signal processing problems can occur because the greater the noise, the less the distance that can be measured. Any average noise level greater than 30 dB is generally cause for concern unless the distance is much shorter than the maximum specified for the transducer and it powers its way through the noise.

Determining the noise source

Noise can be from either electrical interference or from an acoustic obstruction that intervenes and distorts the intended echo. Thus, the first step is to determine the noise source.

Start by disconnecting the transducer from the transceiver. If the measured noise is above 5 dB, go to *Non-transducer noise sources* below. For noise levels below 5 dB, proceed with the following:

1. Connect only the shield wire of the transducer to the controller. If the measured noise is below 5 dB, continue with the next step. If the noise is above 5 dB, go to *Common wiring problems* on page 59.
2. Connect the white and black transducer wires to the controller. Record the average noise.
3. Remove the positive wire of the transducer. Record the average noise.
4. Reconnect the positive wire and remove the negative wire. Record the average noise.

Use the Noise Modification table to determine the appropriate next step. The terms *higher, lower,* and *unchanged* refer to the noise recorded in the previous steps.

Noise modification table

	- removed	+ removed	Go to:
noise	higher	higher	Reducing electrical noise
		unchanged	Common wiring problems
		lower	Reducing acoustic noise
	unchanged	higher	Reducing electrical noise
		unchanged	Contact Siemens
		lower	Reducing acoustic noise
	lower	higher	Common wiring problems
		unchanged	Common wiring problems
		lower	Reducing acoustic noise

Non-transducer noise sources

Because of high costs, filtering cables are not recommended unless all other options have been exhausted.

Remove all input and output cables from the controller individually while monitoring the noise. If removing a cable reduces the noise, that cable may be picking up noise from adjacent electrical equipment. Although controllers are designed to work near heavy industrial equipment such as variable speed drives, do not locate them near high voltage wires or switch gear.

Ensure that low voltage cables are not being run adjacent to high voltage cables or near to electrical noise generators such as variable speed drives.

Moving the electronics a few meters away from the source of noise will fix the problem. Shielding the electronics is another option, albeit expensive and difficult to install properly; the shielding box must enclose the controller electronics completely, and all wires must be brought to the box through grounded metal conduit.

Common wiring problems

The installation wiring can also present noise problems.

- make sure that the transducer shield wire is connected at the electronics end only. Do not ground it at any other location.
- do not connect the transducer shield wire to the white wire. (Applies only to the SITRANS LUT400, MultiRanger 100/200, HydroRanger 200, and SITRANS LUC500)
- the exposed transducer shield wire must be as short as possible.
- connections between the wire supplied with the transducer, and any customer installed extension wire should be done in grounded metal junction boxes.

For SITRANS LUT400, LUC500, MultiRanger 100/200, and Hydro-Ranger 200, the transducer extension wire must be shielded twisted pair. For SITRANS LU series, AiRanger series, HydroRanger Plus, and MiniRanger Plus, use RG-62A/U co-axial cable for noise reduction purposes. See the installation section of the controller manual for correct cable specifications.

NOTE: On Siemens transducers, the white wire is negative and the black wire is positive. If the extension wire is colored differently, make sure that it is wired consistently.

Reducing electrical noise

- ensure that the transducer cable does not run parallel to other cables carrying high voltage or current
- move the transducer cable away from noise generators such as variable speed drives
- put the transducer cable in grounded metal conduit
- filter the noise source

Acoustic noise

To confirm that the problem is acoustical, place several layers of cardboard over the face of the transducer. If the noise is reduced, the noise is definitely acoustical.

Reducing acoustic noise

- move the transducer away from the noise source
- use a stilling well
- install a rubber or foam bushing or gasket between the transducer and the mounting surface
- relocate or insulate the noise source
- change the frequency of the noise. The controller is only sensitive to noise between 25 KHz and 65 KHz

Summary

Sonic intelligence beats at the heart of Siemens ultrasonic technology. Derived from the cumulative experience of over a million ultrasonic applications, this sophisticated echo processing focuses a laser sharp eye on the true level in the application.

Chapter Five

Installation

There is joy in repetition, there is joy in repetition, there is joy in repetition.[1]

Ultrasonic level measurement instrumentation offers non-contacting technology that is virtually maintenance free and cost effective in many applications, from the water and wastewater industries to solids and liquids measurement in a variety of industrial settings.

Ultrasonic systems use a transducer to send electronic pulses to a controller which interprets them and determines level distance. Integrated systems like the SITRANS Probe LU house the transducer and controller together and can also function as transmitters within a network, using the latest in communication hardware and software to transmit data back to its source for immediate processing. They can operate pumps and alarms directly or send data through communication systems.

---

[1] Prince, "Joy In Repetition." *Graffiti Bridge,* 1990.

Other systems separate the transducer from the controller, and only the transducer is actually at the application. The controller can be as far as 365 meters (1200 ft) away. The application can be configured and managed from a distance; thus in dangerous or awkward situations, no one needs to return to the site after the initial installation.

Ultrasonic transducers are very easy to install and maintain, ensuring many years of reliable operation. This chapter discusses installation requirements and best practices to maximize the benefits of an ultrasonic application.

Topics

- transducer selection
- location and obstructions
- vessels
- tanks and access
- nozzles, standpipes, stillpipes
- cabling
- open vessels, flumes, weirs
- lift stations
- position control
- approvals

Installation conditions are the same for the transducer sensors only or for transmitters like the SITRANS Probe LU. The controllers are generally mounted away from the applications where they can be configured in comfort and safety.

Select the right transducer

The Echomax XRS-5 is a popular transducer choice. With a measurement range of 0.3 meters (one ft) to eight meters (26 ft) on liquids and slurries, it is suitable for most short to mid-range applications in the water and wastewater industries.

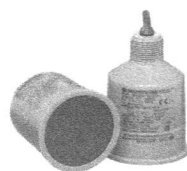

For short-range solids applications, the Echomax XPS-10 provides a high power output suitable for dusty applications.

For applications beyond eight meters (26 ft) and up to 15 meters (50 ft), Echomax XPS-15 transducer offers the most accurate performance for solids, liquids, and slurry applications.

Applications beyond 15 meters (50 ft) require the XPS-30 or XPS-40 transducers with lower frequencies and more powerful output.

Many conditions can affect the performance of your ultrasonic system, and they need to be noted when choosing a transducer. The application environment also influences the performance, and in this chapter we discuss the actual installation criteria for the transducers after you have decided which transducer is appropriate for your application and is compatible with the controller.

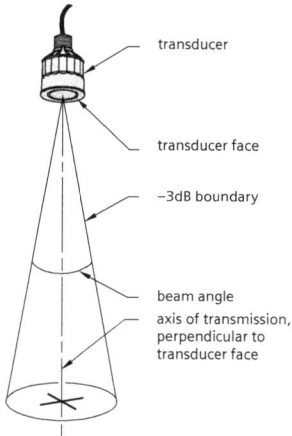

Location

The actual physical location of the transducer is very important for optimal performance, and there are a number of factors that must be considered depending on the application type. Thus the transducer mounting for measuring open channel flow is different from measuring liquids in a vessel and different again from measuring solids. However, regardless of the application type, the one common condition to avoid in any application is the obstruction.

Keep objects out of the ultrasonic cone to reduce false echoes.

Obstructions

Obstructions are the most important application elements to be aware of as they block the sound signal path. Typical vessel obstructions include ladders, pumps, braces, lights, agitators, walkways, and loading chutes. If obstructions are unavoidable, choose the location where there are the least number of obstructions. The transducer must be able to "see" the material being measured, and a beam, pipe, or structural brace will block the view and affect the reading. Obviously, the clearer the access to the material, the better the performance.

Because the ultrasonic pulse radiates in a cone shape from the face of the transducer, keeping objects outside of this signal cone reduces the chance of false echoes being recorded. The transducer beam angle specifies the degree of the cone (see illustration below) where the ultrasonic pulse's power is lower by three dB (decibels).

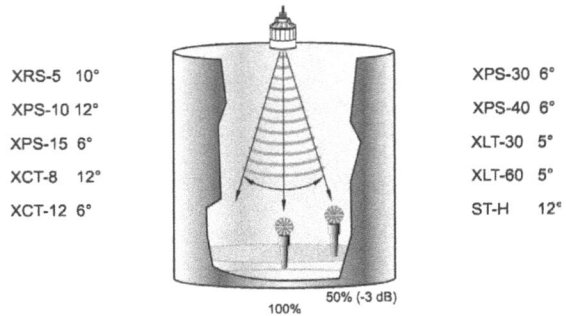

XRS-5 10°	XPS-30 6°
XPS-10 12°	XPS-40 6°
XPS-15 6°	XLT-30 5°
XCT-8 12°	XLT-60 5°
XCT-12 6°	ST-H 12°

100%   50% (-3 dB)

Closed vessels

In closed vessels, locate the transducer on a standpipe or stillpipe, or any access port or nozzle that offers an unobstructed view of the surface to be monitored. The standpipe or stillpipe can negate the blanking requirements by distancing the transducer from the material.

Tanks

The variety of vessels measured by ultrasonics is extensive, from small one meter high (3.3 ft) plastic containers of wood glue resins to three meter high (10 ft) stainless steel tanks full of ice cream base to 60 meter (200 ft) concrete grain silos. Ultrasonic measurement has very few limitations when it comes to vessel size or geometry; it simply needs a clear path for the sound to travel to the material being measured. The shape of the tank matters little.

Although tank or vessel sizes can vary, the majority are variations on only a few standard shapes. Transducer location can change slightly depending upon the shape of the vessel.

These shapes include:

Flat bottom tanks

Common applications include wastewater wet wells. Locate the transducer as close to the center as possible and away from obstructions in the beam path. If mounting in the center is not possible, maintain a 10:1 ratio when mounting close to the side of the vessel.

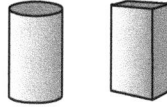

NOTE: For every ten meters vertical measurement, locate the transducer one meter from the side to prevent any interfering echoes.

Cone bottom/parabolic bottom tanks

For tanks used in liquid material applications, mount the transducer in the center so that measurement to the bottom of the cone is possible. If this is not possible, program the controller so that the bottom of the tank is the top of the cone.

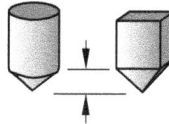

If the material is a solid (i.e. grain), then a transducer with an aiming device should be used and the transducer aimed towards the bottom of the cone to ensure measurement into the draw point.

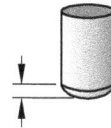

Horizontal tanks

Mount the transducer as close to the top center of the vessel as possible to make use of all the available span.

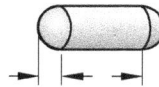

NOTE: Echo processing software can ignore reflections from support beams, weld beads, ladder, or material buildup along the vessel sides.

Tank access

There are many ways to access a sealed tank or vessel:

- cut hole and/or Easy Aimer
- nozzles
- standpipes
- stillpipes
- sidepipes

Cut hole/Easy Aimer: some tanks or vessels have no entry routes, but can be accessed by cutting a hole or using an existing trap door to mount that transducer. The Siemens Milltronics Easy Aimer assists in aiming the transducer through an access port into the vessel at an angle required to get the accurate reading. Attach the transducer to the aiming kit using the coupling provided with the kit and then aim past the transducer so that the sound path is perpendicular to the material surface.

NOTE: Be sure to locate the access port so that no obstructions are in the beam path.

The Easy Aimer aims the transducer to the low level draw point. The transducer can be revolved through 360° and angled at 0 to 27° off vertical. It can withstand 100 kPa (one bar or 15 psi gauge) pressure with aggressive environments.

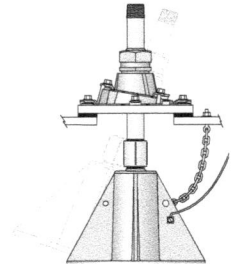

Nozzles: a protruding access pipe located on the surface of the vessel. A short outlet, or inlet, pipe projecting from the end or side of a hollow vessel.

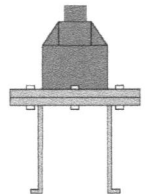

Standpipe: an access pipe that protrudes through a vessel wall and is present on both the outside and the inside of the tank, and may be used as a mounting platform.

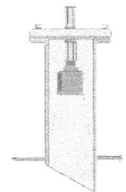

Be aware of the following when using a nozzle or standpipe. When mounting flanged transducers to closed top vessels:

- be sure no buildup forms inside the standpipe (may create false readings)
- eliminate welds and seams as they have the same effect as buildup
- height should be less than three multiples of the diameter (H < 3 D)
- minimum diameter is 76 millimeters (3")
- if height is less than 150 millimeters (6"), factory blanking can be used. If height is greater than 150 millimeters, increase blanking to 150 millimeters beyond the bottom of standpipe, or extend standpipe into tank and cut the end at a 45° angle.

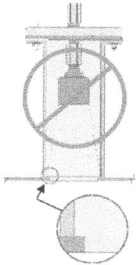

Standpipe installation

Cut the end of the standpipe on a 45° angle and remove any burrs from cutting. This angle will help match the impedance of the air inside the pipe with the rest of the application environment. A 90° cut could create an interfering echo at the end of the standpipe and reduce the confidence value of the real echo.

For standpipe installations, use a factory bonded flanged transducer or split flange kit that will readily mate to the flanged standpipe. Another option is to hang the transducer from a blind flange.

The standpipe should be as short, and the diameter as large, as possible. The -3 dB cone of the sound beam should not intersect the standpipe wall in applications opening into a vessel or larger area. Otherwise, additional blanking will be required to compensate for the interference zone created by the opening.

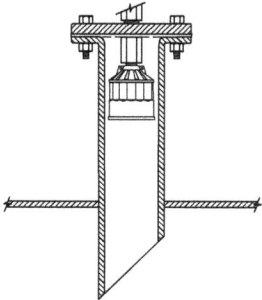

Stillpipes/Stilling tubes

A stillpipe (essentially an extended standpipe) may be used for turbulent liquid applications as it extends further down into the vessel, and is also helpful in foamy conditions by isolating the surface.

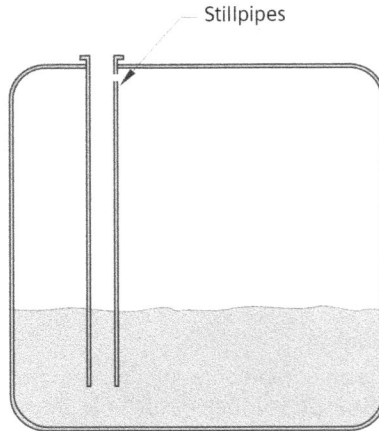
Stillpipes

A stillpipe is simply a smooth walled pipe that extends from the top of the vessel to the bottom. The lower end of the pipe is open to allow the clean free flowing liquid to move up the pipe. In an application where the vessel is emptied completely, it may be possible for the foam to enter the bottom of the stillpipe. In this case, a J-tube is recommended. The J-tube is a stillpipe with an 180 ° elbow at the bottom. When the vessel is emptied there will always be a small amount of liquid in the elbow, and it will not allow any of the foam to pass into the stillpipe.

The same considerations for a standpipe should be applied to stillpipe:

- do not perforate or drill holes along the length of the stilling well
- welds and couplings can affect the reliability of level measurement
- avoid buildup or other debris on the inside of the well wall
- ensure there is a vent hole at the top of the stilling well

Sidepipes

Sidepipes are similar to stillpipes but are on the outside of the vessel and is comparable to a sight glass. The level is reflected in the joined pipe. Sidepipes eliminate false echoes from obstructions and fill streams, and are often used in sumps and wells. The following must be noted:

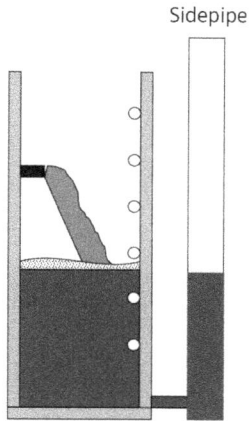

Sidepipe

- pipes should be seamless with a minimum 150 millimeter (6") diameter
- pipe should be vented at the top
- pipe must be kept clean of buildup

Flanges

Transducers can be mounted with a variety of flanges:

- bonded flanges for non-corrosive applications
- PTFE–faced flanges for corrosive applications
- split flanges for field retrofit for mounting on a flanged standpipe

Over tightening the mounting hardware can lead to transducer ringing.

Tighten the flange bolts evenly to ensure a good seal between the mating flanges, but do not over tighten. This can adversely affect performance. Hand tightening should be sufficient.

Liquid applications

For best results in liquid level measurement, the transducer must be perpendicular to the material surface. Whether it is through a standpipe, nozzle, or positioned in the vessel using an existing opening or a newly cut one.

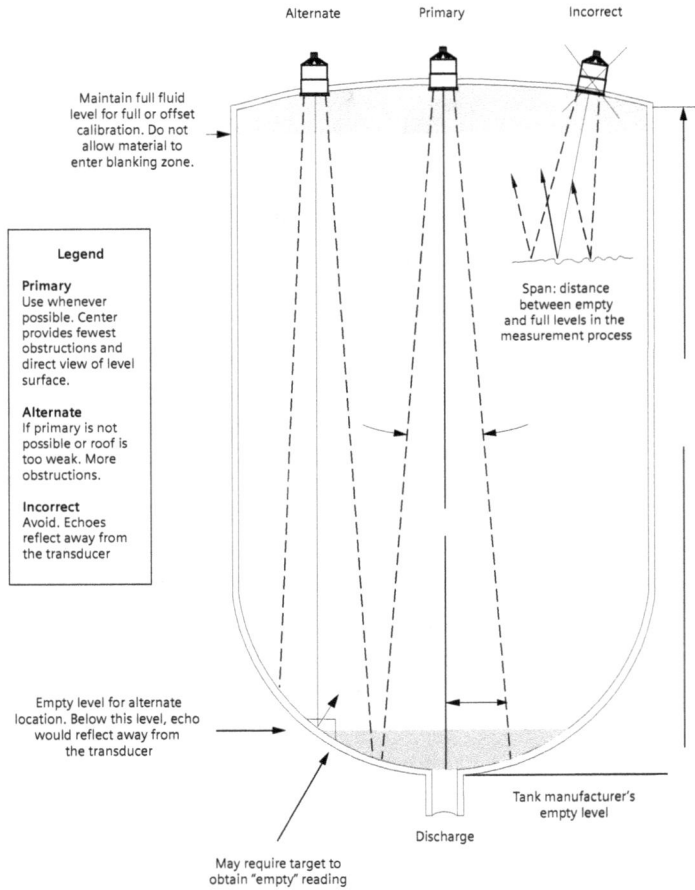

Alternate   Primary   Incorrect

Maintain full fluid level for full or offset calibration. Do not allow material to enter blanking zone.

Legend

Primary
Use whenever possible. Center provides fewest obstructions and direct view of level surface.

Alternate
If primary is not possible or roof is too weak. More obstructions.

Incorrect
Avoid. Echoes reflect away from the transducer

Span: distance between empty and full levels in the measurement process

Empty level for alternate location. Below this level, echo would reflect away from the transducer

Tank manufacturer's empty level

Discharge

May require target to obtain "empty" reading

Solids applications

Note the following when installing the transducer on a tank:

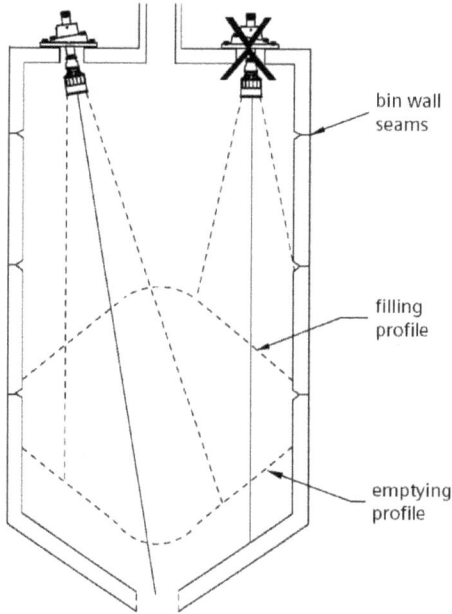

bin wall
seams

filling
profile

emptying
profile

- In most tanks, the transducer should be centered as much as possible (without interference from inlet) for optimum reading range. Whether the tank is empty or full, it must contain its normal vapor conditions.
- Whether the tank is empty or full, it should be at its normal temperature.
- The transducer should be perpendicular to the angle of repose of the material, which usually means mounting it to one side and at an angle.
- In cone bottom tanks, aim the transducer at the draw point at the center of the cone.
- Aim the transducer away from seams, structures and irregularities in the bin wall.
- Locate the transducer away from the material inflow.
- On fluid-like solids, aim the transducer perpendicular to the material surface.

- On dual discharge bins, aim the transducer at the draw or discharge point to ensure accurate readings on emptying. Use Easy Aimer.
- Keep transducer away from fill points, as the material entering the vessel could reflect the sound pulse and give an incorrect level reading.
- Look out for rat holing where the discharge has created a drain hole that can give incorrect level readings. Bridging, where the discharged material does not show, also leads to incorrect level readings.

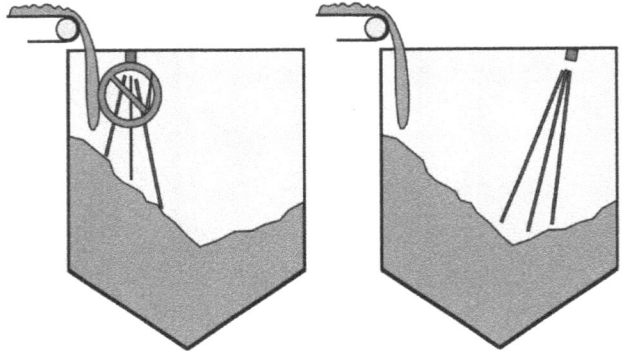

Protection

For solids applications where the transducer can come in contact with material, such as in a rock box or crusher, protect the transducer from flying rocks by surrounding it with a wire cage. Secure the installation by connecting a safety chain from the transducer to the mounting surface.

Use proper cabling

Cable selection is a very important to ensure proper functioning of the transducer. All wiring, conduit, and junction boxes are to be installed according to all local laws, rules, and regulations and/or insurance company requirements. Refer to product manuals for wire type and connections for specific transducer models.

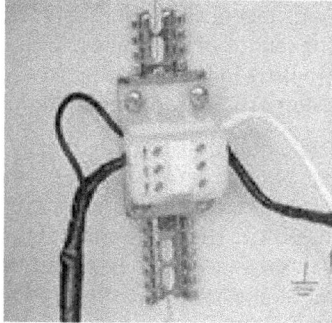

- Install a metal junction box and two conductor terminal strips at an easily accessible location.
- Terminate cables per the transducer installation manual.
- Ensure that the cable shields are isolated from the ground.
- Failure to properly insulate the transducer shield can cause a short to the ground.

- Grounded metal conduit is required between the transducer and the metal junction box.
- Grounded flex conduit is acceptable.
- Rigid metal conduit should be installed from the junction box to the electronics package.
- In the case of multiple point systems like the SITRANS LU10, a single conduit can be used for all transducer cables.
- If transducer cables run in the same conduit or in close proximity to each other, be sure to synchronize the transceivers (see device instruction manual) to prevent cross interference.

Transducers all have integral cables with "pigtail" leads at the end. The length of integral cable can be selected when ordering and are available in lengths of up to 100 meters (328 ft); however, the five meter (16.5 ft) and ten meter (32 ft) lengths are most popular.

The pigtail ends can be directly connected to the controller's terminal blocks, or the cable can be extended using a field-mounted junction box and additional cable, up to a total length of 365 meters (1200 ft). The additional cable type depends on the controller, not the transducer. For extended runs beyond the transducer's integral cable always check the product manual for recommended cable type. Some controllers, like the SITRANS LUT400 and MultiRanger 200, use 18 AWG shielded, twisted pair cable; other models, like the SITRANS LU10, use RG62U coax cable to extend the length of the transducer cable.

All transducer cable should be mounted in grounded metal conduit to provide the best electromagnetic noise shielding and physical protection of the cable itself. Do not run cables near high voltage or current runs, contractors, or SCR control drives.

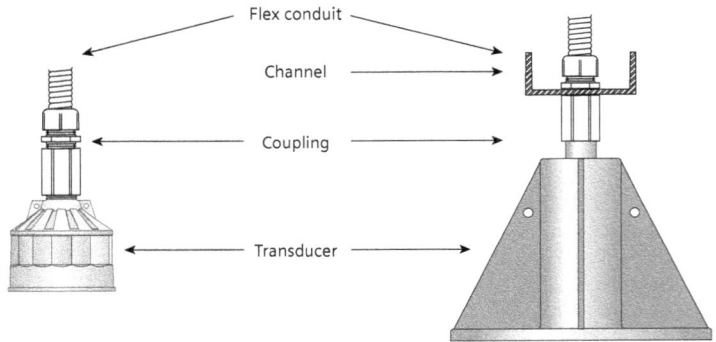

Diagnostic software

For difficult installations involving excessive dust, turbulence, obstructions, or very narrow wells, diagnostic software is available to help fine-tune the transducer's aim and to set parameters. Siemens Milltronics uses SIMATIC PDM for viewing echo profiles and to diagnose echo properties. Support for DTMs and other types of configuration software is available for some models.

Match environment temperature

If a transducer becomes significantly warmer or cooler than the average air temperature in the application environment, it may affect accuracy because the speed of sound changes with temperature. Although some transducers have built-in temperature compensation features, it's always best to mount a transducer where it will share the same air temperature as the rest of the application environment.

For transducers mounted outdoors in direct sunlight, provide a sun shield. If there is a significant temperature difference between the medium and the sensor surface, connect an external temperature sensor to the controller for increased accuracy.

Open vessels

Locate the transducer as close as possible to the maximum height of the material being measured without infringing on the blanking distance. The blanking distance varies among models, so check the product manual for blanking distance instructions related to your transducer.

Mounting brackets

These versatile brackets provide sturdy platforms for open vessels and for open channel monitoring. Manufactured from 304 stainless steel (1.4301), these brackets come in a variety of shapes to fit any application.

Open channel meters: weirs and flumes

Open channel flow occurs when liquid flows in a channel with a free surface open to atmosphere such as streams, rivers, irrigation canals, and ditches. Non-pressurized partially filled pipes, including storm sewers, sanitary sewers, and culverts also qualify. Open channel flow conditions are commonly found in wastewater treatment systems, industrial waste systems, and irrigation systems.

In an open channel, pressure is not transmitted from one end to the other end of the fluid like that in a fully flowing pipe, and only the force of gravity on the fluid causes the open channel flow.

One of the time-proven and common methods of measuring flow uses a primary device to restrict the flow of liquid so that a mathematical relationship exists between the liquid depth (head level) and the flow rate or velocity of the water.

There are two types of primary devices: weirs and flumes.

Weirs

Weirs are the most common type of primary devices used to measure open channel flow. A weir is a dam built across a channel and the liquid flows over it, often there is a cutout or notch, and the weir is generally named for the shape of the notch. A weir is the simplest and oldest type of channel flow measuring device.

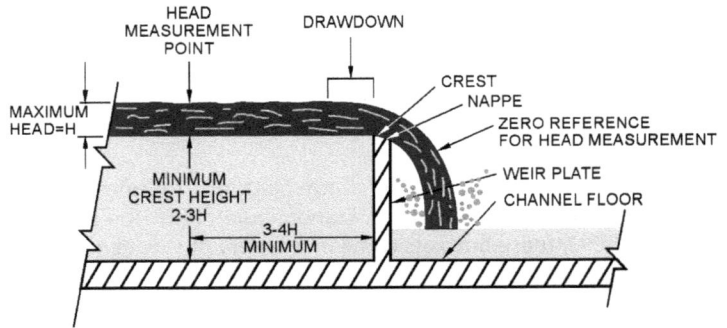

Rectangular weir without end contractions

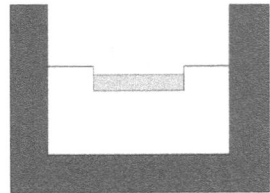

Rectangular weir with end contractions

V-Notch (or triangular) weir

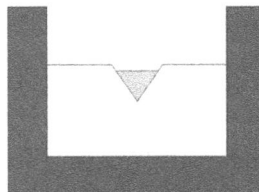

Trapezoidal (or Cipolletti) weir

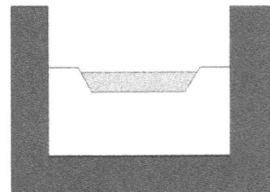

A mathematical relationship can be derived to calculate the discharge through the sharp-crested weirs.

Flumes

A flume is a venturi device[2] placed within a channel to restrict the flow and produce a head (level)/flowrate relationship. The flow induced is between critical and super critical flow levels so only one head measurement is required to measure flow Q.

Most flumes are designed to pass the flows through a critical state for a specified head range, provided flumes are installed, maintained, and operated within the specified range.

Common flumes include:
- Parshall flume
- Rectangular flume
- Palmer-Bowlus flume

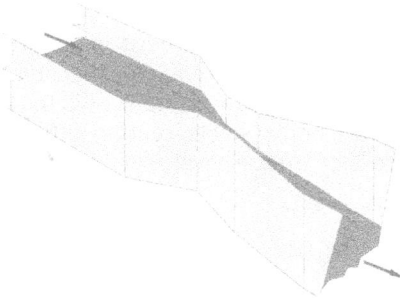

Ultrasonic level measurement devices are ideally suited to measuring water level in primary devices. The non-contacting nature of ultrasonics allows for measurement without disrupting the flow; contact devices used for measuring head level would actually disrupt the laminar flow and affect the head/flow relationship and the overall accuracy of the flow measurement. The surface of the water is smooth (ideally there is no turbulence) and this provides a perfect reflecting surface for sound waves. Ultrasonic devices can also provide integration of the instantaneous flow rate to a totalized amount.

The overall accuracy of open channel flow measurement is dependent on the following:

---

[2] Any device that restricts the flow and regulates it through a measurable point.

- the primary element selection and design
- proper installation of the primary device
- the effect of the approach flow
- the effects of the discharge flow
- the liquid being measured
- the secondary measuring device
- field calibration of the primary element (proving that it does perform as intended)
- field calibration of the secondary device

This section focuses on the secondary device, the ultrasonic open channel meter, and the field-mounted transducer. Many texts and guidelines are available that focus on the primary device and its installation criteria; please refer to these for selecting and installing a flume or weir properly (see *ISCO Handbook*) or www.usbr.gov/pmts/hydraulics_lab/pubs/wmm/cover.html

The primary requirement for accurate measurement is that the approaching flow into a flume or over a weir must be tranquil and non-turbulent. This is perhaps the most important criteria when installing a primary device and also has an effect on the accuracy of the secondary device. If the flow is turbulent in the approach channel then the level will not be stable, causing difficulties for the secondary device to lock onto a level that is representative of the actual flow rate.

Transducer location

The location of the transducer, even if the flow is smooth and tranquil, is important if the head versus flow relationship is to be maintained. For standard weirs, the transducer should be located upstream minimum three times the maximum head level as shown below:

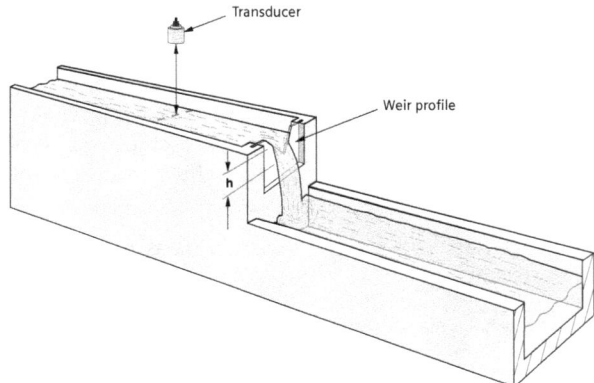

Transducer

Weir profile

h

BS-3680 V-notch weir

ISO 1438/1 (BS-3680) thin plate V-notch weir requires that the transducer be located four to five times the maximum head level back from the weir plate:

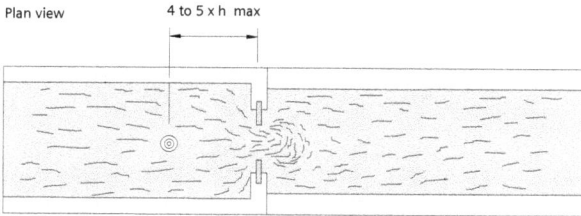

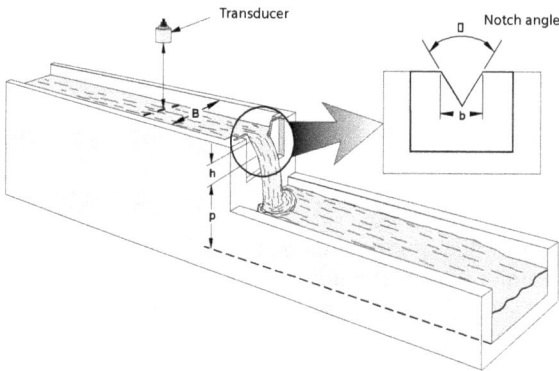

Parshall flumes

For flumes, the location of the transducer depends upon the flume type. Common in North America, Parshall flumes have the head measured at 2/3 the length of the converging section from the beginning of the throat section:

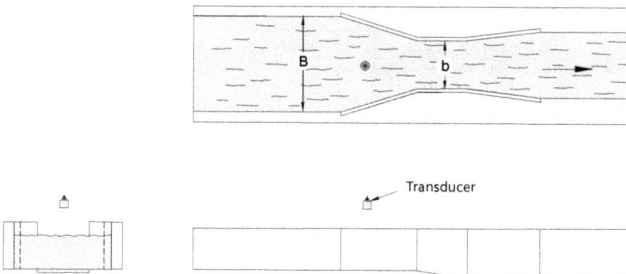

79

BS-3680/ISO 4359 rectangular flume

An ISO 4359 (BS-3680) rectangular flume head level measures four to five times the maximum head level upstream as shown in the following diagram:

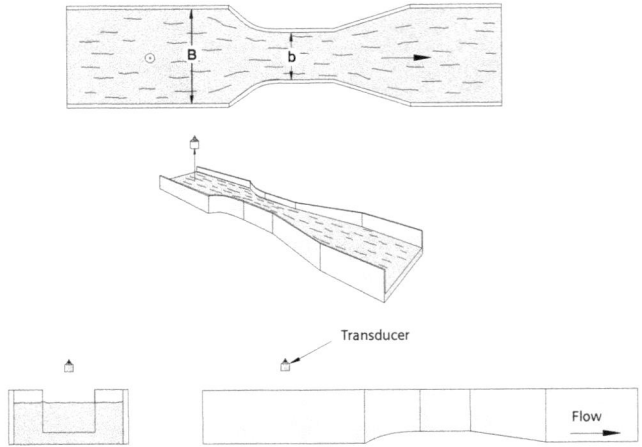

Transducer

Flow

Pipeline flumes

Two types of flumes are designed to fit directly in pipelines in manholes: Palmer-Bowlus flumes and Leopold Lagco flumes, and they are sized according to the pipe diameter. The transducer locations differ between these types; however, both depend upon the pipe diameter. The upstream location of the transducer in a Palmer-Bowlus flume is the pipe diameter 'D' divided by two, as shown:

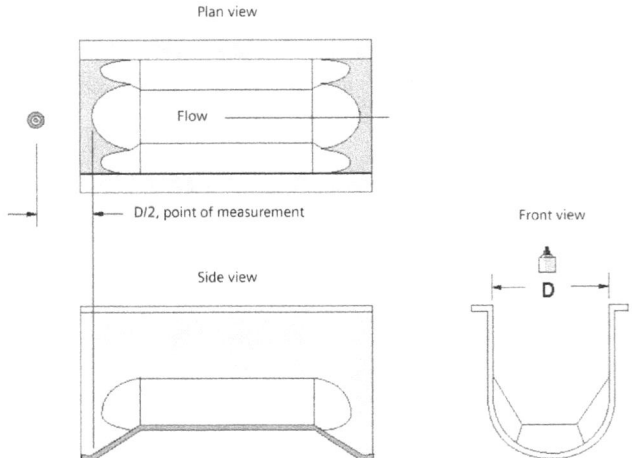

Plan view

Flow

D/2, point of measurement

Side view

Front view

D

The Leopold Lagco Flume head level measurement requires the transducer to be located as shown in the following table:

Flume size (pipe diameter in inches)	Point measurement	
	centimeters	inches
12	2.5	1
15	3.2	1.25
18	4.4	1.75
21	5.1	2
24	6.4	2.5
30	7.6	3
42	8.9	3.5
48	10.2	4
54	11.4	4.5
60	12.7	5
66	14.0	5.5
72	15.2	6

Leopold Lagco flume, transducer location

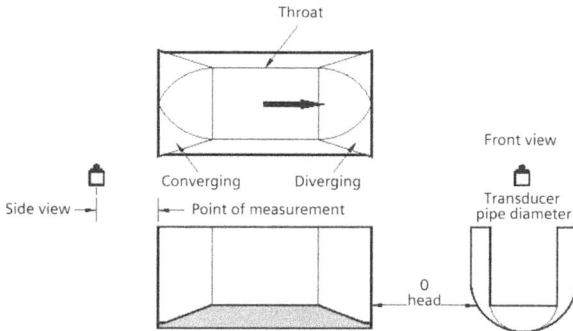

Tips for effective open channel flow measurement

- Ensure the primary device (weir or flume) is the right type and size for the expected flow rate, and ensure it is installed properly. Discrepancies from the standard design, dimensions, or setup will adversely affect the accuracy of flow measurement devices.

- Ensure that it is properly cleaned and maintained. Any buildup of sediment or vegetation can influence flow and flow measurement.
- Check upstream conditions that may create waves or surging. Hydraulic jumps, flow pipes, or drops in approach pipes located close to the flume can interfere with measurement.
- When installing flow measurement devices, follow the installation guidelines in the instruction manual to ensure correct setup and to be aware of any limitations.
- Field calibration of the primary element will tell you if the actual head corresponds to the head-to-flow calibration information supplied by the manufacturer. If not, a new curve can be determined through field testing. This is essential on special applications or non-standard weirs and flumes.
- Install the head measuring device in a relatively calm, stable portion of the channel. It should not be positioned in a channel approach that has high velocity of water, turbulence, gates, valves, pumps, or a sudden change in section.
- Make sure the measurement device has been properly zeroed in reference to the primary device. When the flume is empty, your level sensor should read zero.

Maintenance

If significant performance changes are observed, shut down the level measurement system and inspect it, starting with the transducer. Ensure the transducer is still firmly mounted, wipe the face if it has become soiled, and check the wiring conduit to ensure it is all sound. If the transducer works properly, check the vessel sides for new obstructions or material buildup.

If the sound path is clear, perform diagnostics at the controller. Troubleshooting information is clearly marked in the instruction manual, and other procedures include analyzing the echo profile for echo quality (strength, confidence, and noise).

Generally, low maintenance is a key advantage of ultrasonic transducers. The powerful pulsing action makes the face self-cleaning. Because transducers are non-contacting, there is no material buildup, avoiding the nuisance, danger, and expense involved in cleaning, adjusting, and repairing contacting devices.

Lift stations

Lift stations, also known as pump stations or wet wells, are essentially vertically mounted, flat bottom vessels that happen to be built into the ground. They are part of gravity fed sewage collection systems and are used to collect sewage at one point and then pump it (or lift it) to a higher elevation to eventually end up at the wastewater treatment facility.

Lift stations are also found at the inlet to the wastewater treatment facility itself. The wet well differs from a simple vertically mounted flat bottom vessel by the presence of equipment inside the well: pumps, chains, inlet and discharge pipes, back-up floats, power lines to the pumps, and ladder rungs. Lift station environments are also very corrosive and dirty, and the water can contain solids; thus contact devices, like floats, can prove to be unreliable over time as material buildup and constant contact wears on these devices. Ultrasonic transducers, a non-contacting technology, are mounted above the sewage and only come in contact with sewage during abnormal flood conditions. Thus ultrasonic level measurement provides near maintenance-free operation for many years.

Install the transducer on a sturdy bracket or hang from metal conduit. The mounting height should at least 30 centimeters (12") above the expected maximum level; locate the transducer as far away as possible from the inlet pipe, pumps, ladder rungs, floats, the wet well wall, and other obstructions. If extremely turbulent or foamy, use a stilling well and follow the installation guidelines outlined on page 68.

Wet well transducer selection:

- short range wet wells [0 to 8 m (0 to 26 ft)]: XRS-5
- short range wells with obstructed beam path: XPS-15
- deeper wells: XPS-15
- very narrow wet wells: XPS-30

Flood alert

If the transducer is submerged, especially in applications such as a flooded wet well, use a transducer with a submergence shield that maintains an air pocket in front of the transducer. The controller recognizes the unique echo created by the air pocket, and the controller outputs a high level reading.

When installing a submergence shield, ensure it is properly seated and even across the transducer. Check for leaks by placing the assembled unit in a bucket of water and see if any bubbles leak out. Leaks indicate the shield has not been pulled down far enough to properly seat the pressure clips.

NOTE: As you select the transducer, you also need to select a compatible controller to process the signal from the transducer to provide the following:

- level monitoring
- data logging
- advanced echo-processing technology
- pump control functions
- remote monitoring

All controller models offer backlit displays for easy readability and other advanced features that can improve plant efficiency. Consider your specific application needs against the many features that are available.

Position control

Ultrasonic sensing technology has other uses that are outside of level measurement but that operate on similar principles. The transducers act as distance sensors, indicating proximity and activating alarms or motors or switches. This ultrasonic positioning control requires the transducer to be mounted horizontally and is very effective in the shipping industry for operating overhead cargo cranes and in the mining industry for managing coal loading conveyors.

Hazardous approvals

Transducer and transmitters are installed in potentially explosive environments. Great care and effort is taken during the development of transducers to ensure they are safe for use and meet or exceed the approval requirements necessary for installation in these environments. Not only must they perform to the distance and temperature ranges specified, they must also do so safely in hazardous or explosive environments.

Under most governing electrical codes, it is up to the user to ensure that the transducer or transmitter being installed has the necessary approvals to meet the minimum area classification of the application. Transducers from Siemens Milltronics come with a wide range of approvals to meet the demands of the applications that have explosive environments. They vary slightly from country to country as the world has not adopted a scheme for global approvals.

Approvals

Siemens ultrasonic transducers are designed to meet the requirements to be used in both gas and dust atmospheres and are certified by various agencies around the world. It is important when specifying and installing any equipment in a location that is classified as hazardous to follow all local electrical codes and guidelines and to pay particular attention to the specified ratings.

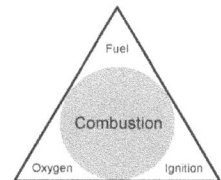

NOTES: The fire triangle or combustion triangle illustrates the conditions required for most fires. Three elements are needed: heat or spark, fuel, and an oxidizing agent (usually oxygen). Fire is prevented or extinguished when any of the conditions are removed.

In the world of electrical engineering, a hazardous location is when concentrations of flammable gases, vapors, or dusts occur. Electrical equipment installed in these locations is especially designed and tested to ensure it does not initiate an explosion through arcing contacts or high surface temperature.

Controller installation

While the transducer/compound instrument installation must be carefully planned because of its proximity to the application, the controller installation can be determined by the operator's requirements. Thus, if the operator is near the application, the controller can be placed close by. However, because the controller can be installed as far away as 365 meters (1200 ft), it can also be placed indoors and away from the elements where it may be installed independently or in a rack as one of a number of instruments.

Whether the unit is wall mounted or installed in a rack or panel with other instruments does not affect performance. They are the same, regardless of mounting designs.

Mounting instructions

The wall mount and panel mount units install differently and the manuals contain the specific instructions.

Wall mounted units must be mounted directly to a wall or to electrical cabinet back panel. Alternate surfaces must be strong enough to support four times the weight of the unit.

Rack and panel mounted units will fit into existing slots or, as with the SITRANS LUT400, will require making a cutout to accommodate the unit. A full size cutout template is provided with your unit or may be downloaded from www.siemens.com/level.

Wiring

Cable is brought in either directly through the cable glands or through conduit, depending on the installation requirements:

Cable glands: if the purchased unit does not have pre-drilled cable gland holes or knockout, you need to drill the entry holes. Make sure you remove the mainboard, being careful not to damage the electronics with static electricity. Make sure cable gland holes do not interfere with the lower areas on the terminal block, circuit board, or communication card.

After loosely attaching the glands to the enclosure, thread the cables through, keeping the power cable separate from the signal cables. Then tighten gland to a ensure waterproof seal when wiring is completed.

Conduit: you need to drill the entry holes or remove the knockouts, so make sure you remove the main board, being careful not to damage the electronics with static electricity. Make sure conduit holes do not interfere with the lower areas on the terminal block, circuit board, or communications card. Attach the conduit to the enclosure using only approved suitable size hubs for watertight application. Reinstall the mainboard with the mounting screws.

NOTES:
- Perform installation only by qualified personnel and in accordance with local governing regulations
- Follow proper grounding procedures. This product is susceptible to electrostatic shock. All field wiring must have insulation suitable for at least 250 V. Hazardous voltage is present on transducer terminals during operation
- Supply AC and DC terminals from approved sources
- Use grounding type bushings and jumpers so the non-metallic enclosure does not provide grounding between conduit connections
- Ensure ambient temperature is always within its approved range. Typically -20 to 50 °C (-5 to 122 °F)
- Mount unit so display window is at shoulder level and easy access for programming is provided. Make sure mounting surface is free from vibration
- Leave sufficient room to swing unit lid open for clear access
- Avoid exposure to direct sunlight (Provide a sun shield to avoid direct sunlight.)
- Avoid proximity to high voltage/current runs, contacts, SCR, or variable frequency motor speed controllers

Summary

Ultrasonics is a versatile and friendly technology for many level applications, but the installation requirements must be met for optimal performance. Situating the sensor or transducer in the proper vessel location, following all the safety regulations and approval requirements, and accounting for temperature and pressure conditions will all ensure reliable, accurate performance. Always account for the conditions when choosing a level technology, and those which are more problematic for ultrasonics – dust, vacuum, interference – can be handled by the SITRANS radar line.

Chapter Six
Applications

I'm a cosmopolitan sophisticate of culture and intelligence.
The culmination of technology and civilized experience.[1]

Ultrasonic level instruments are not exclusive to any one industry and are widely used across many businesses and applications. However, there are certain applications where Siemens Milltronics ultrasonic level instruments operate particularly effectively and accurately. Across the globe and over a million strong, these ultrasonic level devices are found in the water/wastewater (W/WW) industry; in mining, aggregate, and cement (MAC) applications; and in the chemical markets.

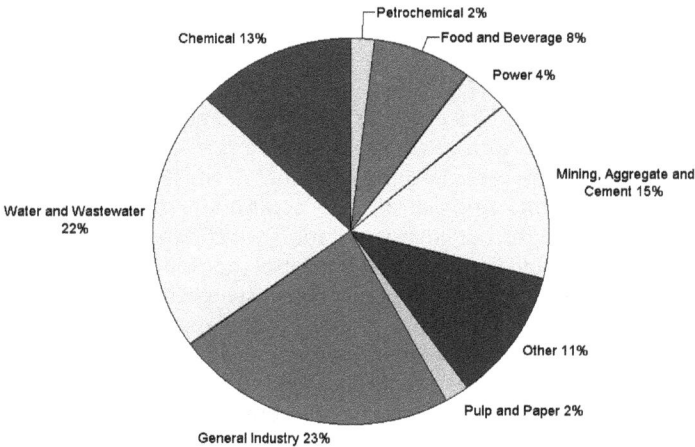

Regardless of the type of application, however, the primary requirement for all industries is to be able to measure level accurately, consistently, and cost effectively. Ultrasonic level measurement meets these needs and has become the selected solution for many industries since the 1970s.

---

[1] Billy Joel, "Running on Ice." *The Bridge*, 1986.

Applications

Markets like water/wastewater and mining, aggregate, and cement employ level measurement for many different purposes, including process control and inventory monitoring. Each industry has a rich history of instrumentation development. As competition grows and resources are strained, controlling processes and inventories is increasingly crucial to the bottom line. Accuracy and reliability become paramount and closer attention is paid to the instruments used to measure inventory and control processes. Measuring and controlling the level of materials contained in storage and process vessels, such as tanks, wells, reservoirs, ponds, bins, and hoppers is one of the most common procedures of industrial instrumentation.

In all these applications, ultrasonics provides two types of level measurement:

- point measurement
- continuous

The point level can also be combined with the continuous level to provide safety redundancy.

Not that long ago, bins and silos were measured by knotted ropes or by sight lines, with employees climbing to the top of the container or in some cases into the container itself. This type of measurement became impractical, and occasionally dangerous, and companies sought out instrumentation which was accurate and reliable, and could be monitored in control rooms or activated from remote locations. Plumb bobs and contact switches could not stand up to the strenuous demands of many of the applications, so non-contacting ultrasonic level measurement as developed by Siemens Milltronics in the 1970s and 1980s was introduced to the market. The rugged reliability of Siemens Milltronics instruments like The Probe and the MultiRanger quickly proved itself invaluable for the following:

- reduce spillage (overflow) – environmental, loss control, legislation
- cost accounting – inventory
- quality control – process control by level
- reduction of inventory in process – smaller vessels
- pump protection – protect from running dry

Topics

This chapter discusses these applications, focusing on the following industries:

- cement
- mining
- water/wastewater
- food
- chemical

Cement

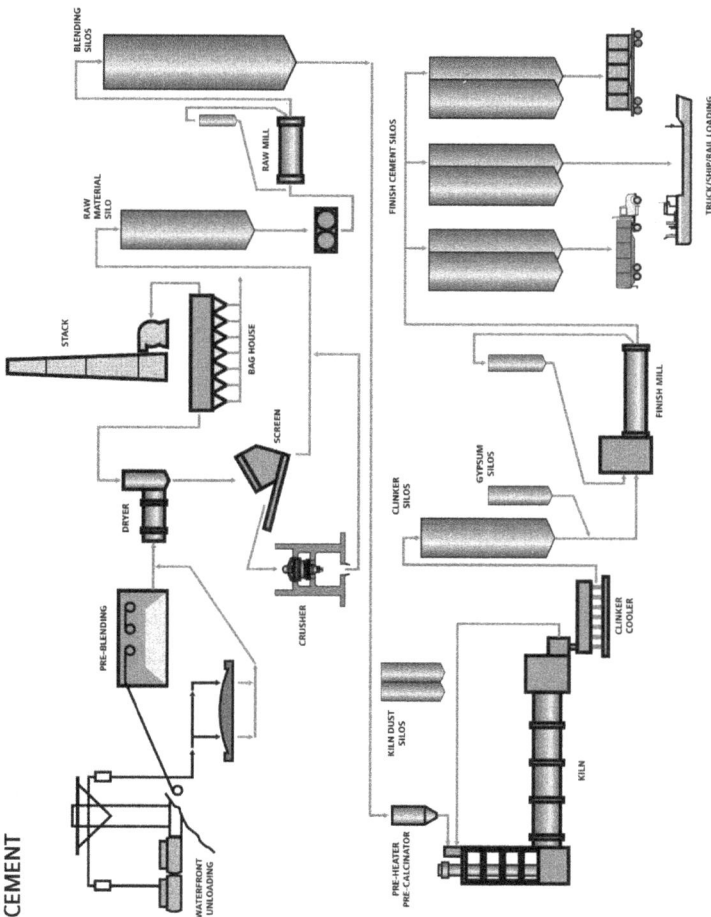

CEMENT

BLENDING SILOS

RAW MILL

RAW MATERIAL SILO

STACK

BAG HOUSE

SCREEN

DRYER

PRE-BLENDING

CRUSHER

WATERFRONT UNLOADING

FINISH CEMENT SILOS

TRUCK/SHIP/RAIL LOADING

GYPSUM SILOS

CLINKER SILOS

FINISH MILL

KILN DUST SILOS

CLINKER COOLER

KILN

PRE-HEATER PRE-CALCINATOR

The cement industry, similar to mining, relied on rudimentary level measurement systems for many years. Plumb bobs or yo-yos, tilt switches, and manual dips provided most of the inventory and process information. In the 1970s and 1980s, ultrasonic measurement appeared on the scene with some of the early Milltronics AiRangers and LR transducers, finding many uses in the cement industry.

The production of finished cement requires numerous level instruments along the way, operating in two capacities:

- as process controllers, activating pumps, belts, alarms, and other actions
- as inventory managers, monitoring material levels in silos and bins

The cement process chart on page 91 indicates the many areas where ultrasonic instruments operate in the cement production process. Other Siemens Milltronics equipment, like belt scales, weighfeeders, and process protection instruments also play vital roles in the production of cement, reinforcing Siemens' role as a total solution provider.

Primary crushers

Gyratory cone crushers are the most common, and there are generally two main applications at the primary crusher.

- *Truck positioning* – the transducer is horizontally mounted in the path of the reversing truck and is wired into a circuit which controls red and green lights. As the truck approaches, the ultrasonic system continuously monitors its position. When it is the proper position for dumping, an alarm relay closes and the green light is activated, signaling to the driver that the load may be safely dumped into the crusher.

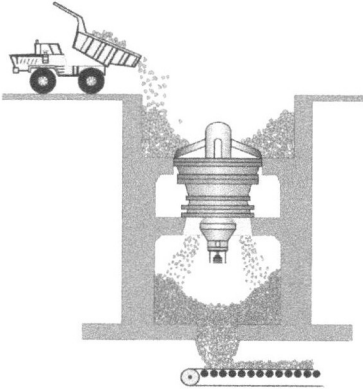

- *Rock box* – located below the crusher, the ultrasonic system monitors the level of the crushed rock. Alarms are set to indicate a clog in the out feed and to indicate when a load can be safely dumped into the crusher. A low level alarm can be programmed to maintain a protective layer of ore in the rock box to prevent feeder damage.

Quarry stock piles

The crushed limestone is transported from the crusher to a stockpile by a stacker conveyor. The ultrasonic transducer monitors stackout profiles and provides inventory estimates in stockpiles to ensure uniform stackout.

An ultrasonic transducer mounted on the head of the conveyor monitors the distance from the discharge to the top of the stockpile. When the distance between the two reaches the programmed set point, an alarm triggers, signaling the need to raise or rotate the conveyor to a new position. This inventory management is a common application for ultrasonics systems where the rugged design and remote positioning of the controller make it an ideal technology for these conditions.

Secondary crusher

In these typically noisy, wet, and dirty applications, ultrasonic transducers sense the level of material in the surge bins. Ultrasonic systems are also used at in-feed chutes to crushers, monitoring for plugged chutes.

Here, the Pointek ULS 200 monitors the chute at the end of a conveyor. If the chute plugs, a relay in the ULS 200 closes immediately, notifying the user that a problem exists and shutting down the conveyor.

Raw mill feed silos

Crushed ore is transported from the secondary crushers to the raw mill feed storage silo. Here an ultrasonic system monitors the level of ore in the silo.

This application is strictly inventory management, as the ultrasonic system monitors the content levels only.

Raw mill

The raw mill is typically a roller mill that pulverizes raw material, including additives, into a fine powder which is then transported to the kiln feed. Blended raw materials are continuously fed into the raw mill.

The ultrasonic system monitors the height of the mill rollers above the table, essentially monitoring bed depth. Operators integrate material feed rate with the roller height to maintain a consistent bed depth, improving milling efficiency and preventing the rollers from striking the mill table.

Kiln feed silo

Crushed material from the raw mill is transported to the kiln feed silo (sometimes called Continuous Feed Silo as material is constantly fed in and out of this silo) where it is stored and preheated in preparation for delivery into the kiln itself. Ultrasonics monitors the level within these silos, but with only a limited range due to the extremely dusty and turbulent environment.

In recent years, many kiln feed silos have been outfitted with Siemens SITRANS LR560 radar level transmitters with great success at overcoming the harsh environments. Ultrasonics may work, however, depending on the dust levels, and is a cost-effective alternative to radar in appropriate applications.

Clinker storage

The material leaving the hot kiln is called clinker. Ranging in size from dust to fist sized clumps, the clinker is stored in silos before being ground into a fine powder. Because it has come out of the kiln, the clinker is quite hot.

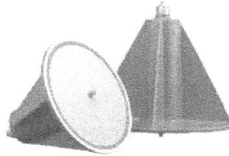

To work effectively, powerful high temperature transducers like the XLT-60 measure the level in the clinker silos. Because clinker applications, although dusty, are not as dusty as kiln feed silos or finished cement silos, ultrasonics is a viable choice for this application; however, more users are switching to SITRANS LR560 radar because of its improved performance.

Additive bins and coal storage silos

These bins contain materials that are added to the raw mill process and are mixed with the crushed limestone before the kiln. They also hold the materials that are added during the final grinding of the clinker. These bins contain sand, iron ore, fly ash, gypsum, plaster, or other materials used to adjust the composition of the cement.

The coal silos store the coal that is burned in the kiln – the heat from the burning coal is responsible for the chemical changes of the raw feed into clinker. Ultrasonic systems monitor the level in these bins for inventory purposes.

Finished cement silos

These silos contain finished cement. Once the clinker and other additives have gone through the ball mills, the material is reduced to a very fine powder and is conveyed to storage silos where it waits for bulk shipping via train or trucking, or for further production into bagged, finished cement.

Finished cement is extremely fine and there is very little moisture, which makes this application very dusty. The performance of ultrasonic systems on this application is inconsistent – it depends on many factors including silo height, dust collection system, transducer location, transducer aiming, and method of filling. Like the kiln feed silos, this application was traditionally monitored by an ultrasonic level system. In cases where the dust is too great, a SITRANS LR560 radar level transmitter operates with great success.

Mining

Inventory management and process control in the mining industry was a rough business. The nature of the product made it difficult to assess volume accurately and most of the control was done manually. Mechanical measurement relied on yo-yos, or occasionally nuclear measurement for point level control. However, nuclear technology is very expensive and has its own safety concerns and yo-yos are prone to breakdown. Processes were also visually monitored but often broke down when the monitor was not watching or away from his post. Crushers jammed up from too much load or ran empty when bins drained early. Efficiency was thus difficult to maintain. Safety was also a problem as workers often had to enter storage bins to check levels or to repair jammed equipment.

The implementation of ultrasonic measurement has greatly improved operational efficiency and safety in mining. The non-contacting technology allows for inventory and process management from the safety of a control booth. Crushers and feed belts operate without interruption and loads are accurately and efficiently processed.

Ultrasonic level measurement operates in various locations throughout mining. Some of them, like the primary and secondary crushers, are identical to those used in the cement industry. Other applications, as shown in the process diagram below, require the use of ultrasonic level equipment for process control and inventory management.

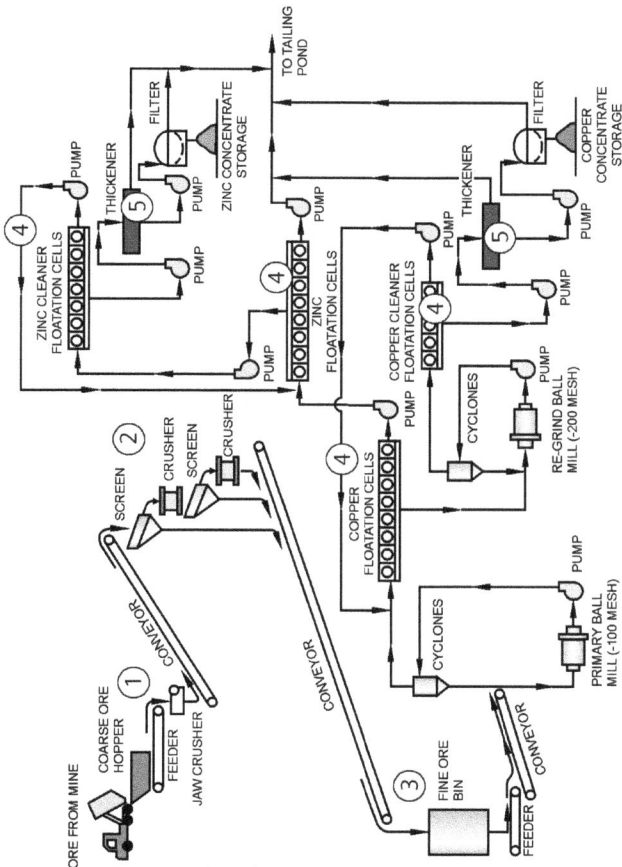

① Primary crusher

② Tertiary crusher

③ Fine ore bins

④ Flotation cells

⑤ Disk filters

Crusher

Ore levels in surge bins are measured by ultrasonic systems to monitor the feed into a crusher. The transducers, although rugged industrial devices, must be mounted far enough away from the feed to reduce repeated collisions from the ore itself.

The crusher reduces the ore to a much finer consistency so it can be added to another process.

Fine ore bins

The XPS transducer series are ideal for measuring fine ore levels, a typically dusty environment. The multi-point SITRANS LU10 controller can monitor level in the bins and can be programmed for priority scanning to monitor bins that are filling at a higher rate than static bins.

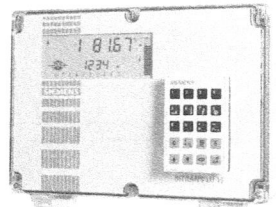

Flotation cells

The flotation process used in mineral processing begins by injecting a frothing agent into mineral slurry, then it pumps the material into an agitated open tank or an aerated flotation cell.

The valuable minerals cling to the froth, which is then scraped from the top of the slurry, after which the frothing agent is removed from the mineral-rich concentrate. The concentrates are dewatered and the waste is discarded as tailings. For an efficient extraction process, it is essential to monitor the level of the slurry and the level of froth.

During the flotation process both the height and thickness of the froth must be calculated to determine the dosage of the expensive chemical reagents. Accuracy is important as it saves money by using only the required amount of reagents and by removing the maximum amount of minerals from the froth.

Two transducers or SITRANS Probe LUs are mounted over the flotation cell. The first measures the level of the froth; the second measures the level of the pulp. The pulp level is determined by using a float ball/target plate assembly: the float ball floats on top of the pulp, a rod with a target plate is attached to the ball and the transducer monitors the level of the target plate. The level of the target plate is directly proportional to the level of the pulp. The signals from the transducers are sent to a process control system which calculates the froth thickness and controls the dosage of the chemical reagents accordingly.

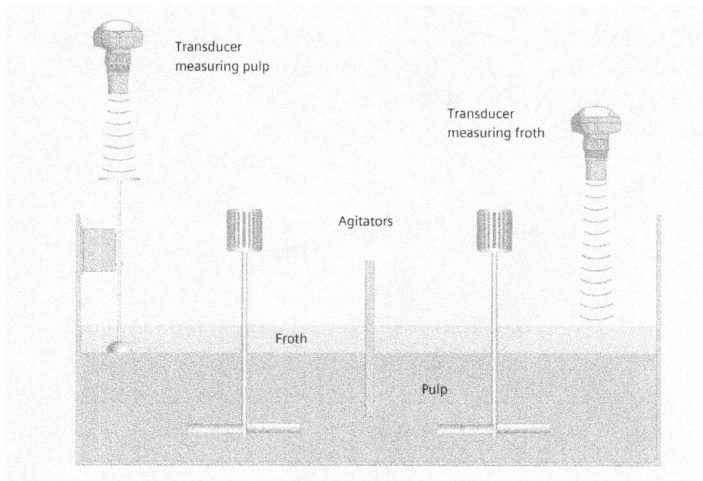

Chemical storage

Many chemicals are used as part of mineral processing and are stored onsite. Ultrasonic systems are used for inventory monitoring of these chemicals. Non-contacting measurement is ideal for this type of inventory monitoring, as the aggressive nature of many chemicals make contact dangerous or impossible.

Slurry tanks

Short range ultrasonic transducers are flange-mounted above slurry tanks to measure the amount of liquid. The PVDF enclosures protect the transducers from most chemicals found in slurry tanks.

Sump pump wells

Many sump wells at a mining site can be measured by short range ultrasonic transducers for more reliable performance, especially when compared to simple floats or other contact switches. The management of water, whether it be to control flooding or as part of the refinement process, is crucial to the mine operation.

Ore storage bins or stockpiles

Often ore is stored in bins or stockpiles between primary and secondary crushing stages. Ultrasonic systems monitor level in the bins or stockpiles. This application is similar to the quarry stockpiling of material used for cement production.

Disk filters

After the concentrate has passed from the flotation cells and the thickeners, disk filters dewater the concentrate before it proceeds to the smelter processing. Ultrasonic systems monitor the level within the filters and activate alarms if the filtered cake (the dewatered concentrate) backs up, thus preventing a spill.

Application one: measuring level in flotation cells

One of the world's leading producers of copper and molybdenum operates a copper mine in the western USA. The operation uses a series of cone crushers and ball mills to reduce copper ore to very fine particles which are pulped (mixed with water) to create a slurry. Small amounts of special reagents are added, including a frother and a collector chemical that causes the copper minerals to stick to the bubbles. This material is pumped to a series of froth flotation cells to concentrate the copper where air is blown into the flotation cell tank, and the pulp mixture is vigorously agitated like in a high-speed blender. Rising bubbles carry the copper particles up and over the edge of the flotation tank where the bubbles break soon after they flow over the edge. The copper is then ground even finer and purified by additional flotation processes.

Conditions

- need to measure the level of both the pulp and the froth to ensure optimum copper concentrate production. Those levels provide feedback to the plant's process control system so operators can make appropriate changes to variables like pulp and air flow as well as the addition of chemical reagents
- difficult for level measurement because of material build-up and plugging that occurs on most types of level technologies
- narrow measurement requirements: span for the pulp measurement (using the Probe LU with target and float assembly) is 25.5 centimeters (10") and the froth level Probe uses a span of only five centimeters (2")
- harsh conditions require a very robust device

Solution

The SITRANS Probe LU ultrasonic level monitor combines sensor and electronics in a single package for effective liquid level measurement in open or closed vessels. The patented Sonic Intelligence echo-processing filters out false echoes from acoustical or electrical noises and agitator blades in motion, giving superior measurement reliability.

Each flotation cell is equipped with a Probe LU to monitor the froth level while a second Probe LU measures the pulp level based on a metal target attached to a mechanical float assembly. The 4 to 20 mA analog output from the Probe LU feeds flotation cell level information into the plant's control system.

Benefits

The SITRANS Probe LUs are very easy to program and they require no maintenance. The only setup required is programming the span for the analog output which is readily accomplished in the instrument shop or during installation.

Aggregate

The aggregate industry is similar to cement and mining in that it involves many of the same processes and much of the same material. The difference is that it occupies the medium stage, and while there are some processes involved (see diagram on page 103), much of the requirement is for inventory management and transfer control. Applications requiring primary and secondary crushers, as well as ore storage bins or stockpiles, are very similar, if not identical, to the applications in the cement industry. Others, like the tertiary crusher, are the same as those found in the mining industry. There are some applications unique to the aggregate industry, and these are shown in the process diagram.

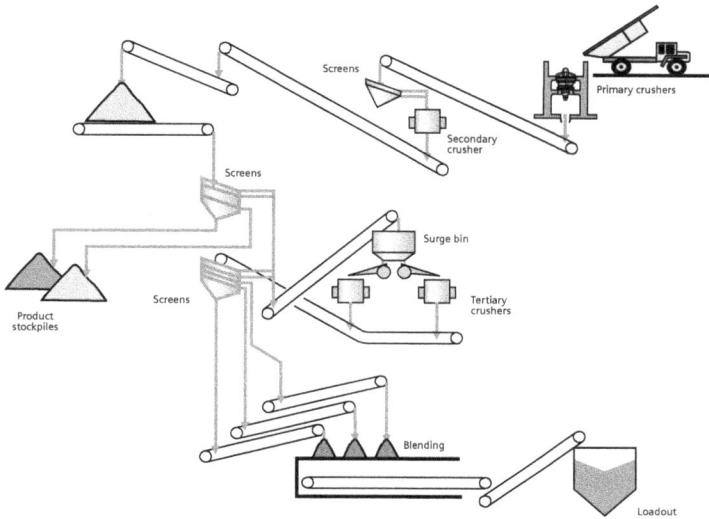

Blending silos and storage bunkers

These large bunkers store the crushed ore. Long range transducers with Easy Aimers provide maximum performance. Loading is often done by tripper cars, unmanned devices that move along a track over the bunkers. The tripper cars position themselves over empty bunkers and load the ore.

Screening station

At the screening station, a multipoint ultrasonic level device like the SITRANS LU10 measures the individual inventory of bins based on material size.

103

Summary

Ultrasonic level instruments play an important role in the Mining, Aggregate, and Cement (MAC) industries. The rugged, non-contacting technology is cost-effective, accurate, and reliable; and can be used for process control, inventory management, and even as distance locators for crane operators.

Environmental

The environmental industry is the biggest market for ultrasonic sensors. From water/wastewater systems to clean water systems, sensors measure flow, run pumps, activate alarms, and control additive inventories.

Collection system: lift station/pump station/wet well

In municipalities and cities with water treatment facilities, an underground, gravity-fed collection system transports raw sewage from residential and industrial users to the sewage treatment plant. Pump stations (aka: lift stations) are distributed throughout the system in places where gravity needs a helping hand. Inside the pump station, sewage is collected in the wet well, a transition point from which it is pumped out, or lifted, until it reaches the treatment facility.

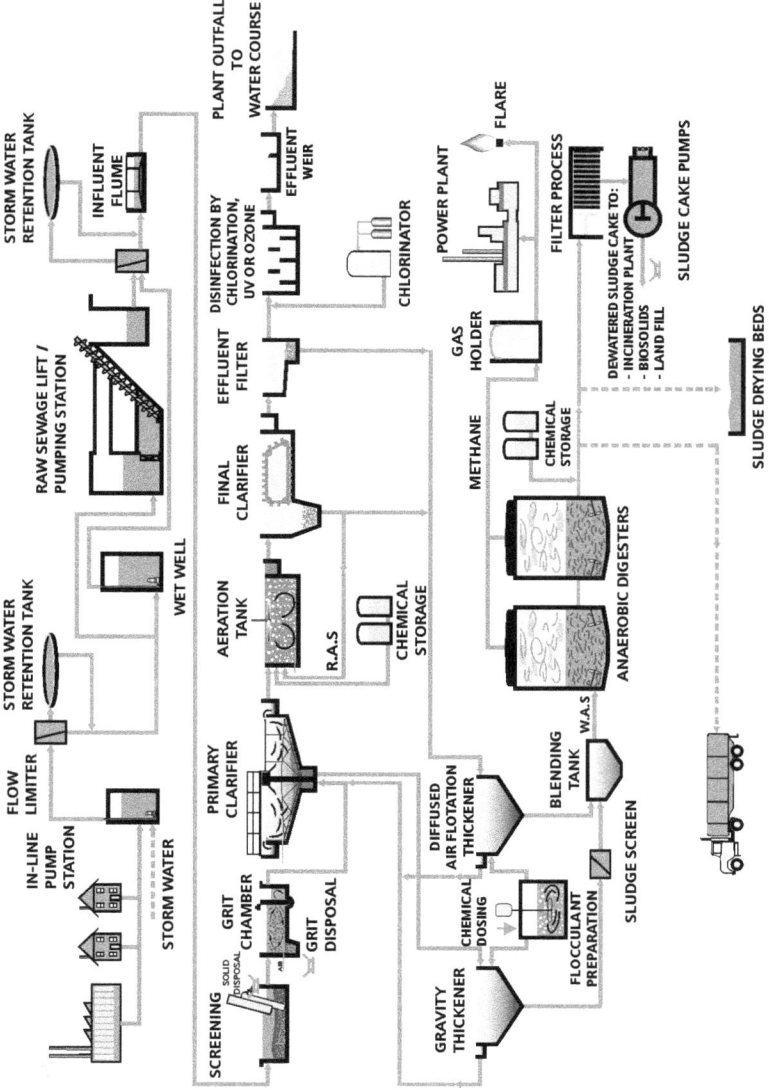

TYPICAL ACTIVATED SLUDGE TREATMENT PLANT

TYPICAL CLEAN WATER TREATMENT PROCESS

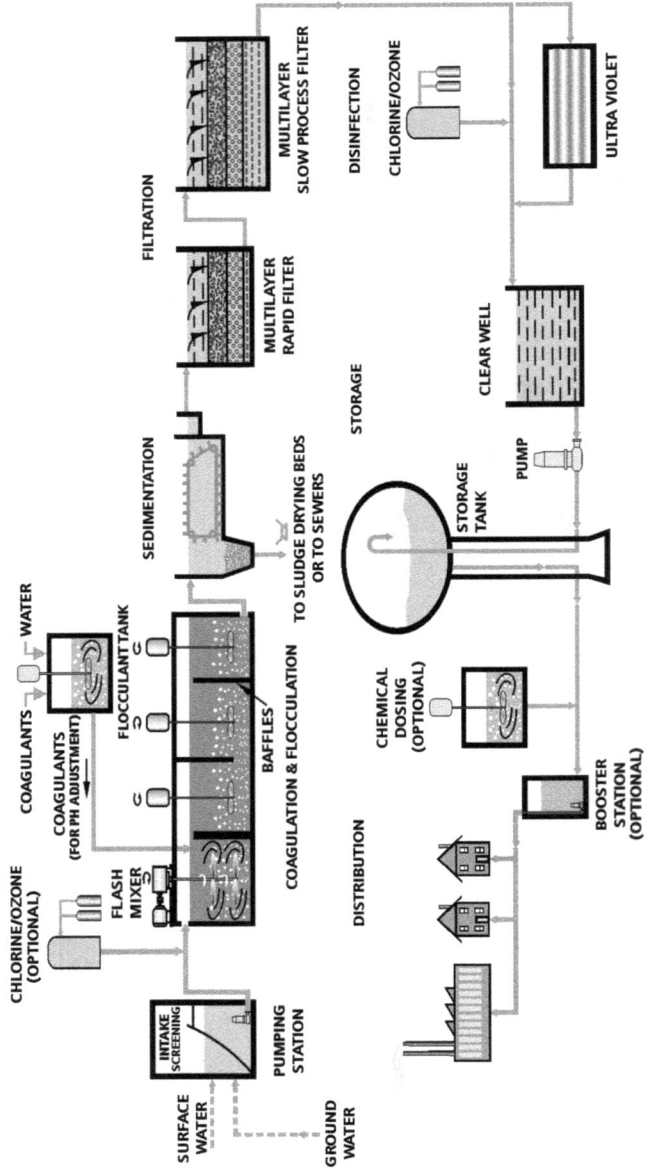

Ultrasonic systems monitor and control the level of sewage within the lift station. As you can see from the photos on page 104 of the inside of a lift station, there are many obstructions in the wells – the narrow beam angle and the non-contacting nature of ultrasonics make it an ideal technology for monitoring wet well levels.

Advanced features incorporated into Siemens Milltronics controllers provide more than simple pump control. The SITRANS LUT400 can monitor pump run time, pump faults, pump starts, and rotate pumps to even the wear on the pumps or wear one pump preferentially.

The LUT400 has unique timing features that add up to great benefits. Pump ON/OFF setpoints can be changed automatically based on the time of day, avoiding pumping during peak electricity price periods and generating energy savings. The LUT400 real time clock control allows pump set points to alter according to the time of day. In a typical wet well installation, this could mean pumping out the well before peak rate and then deferring pumping until after the expensive period. Penalties imposed by electricity suppliers for usage during peak hours can be avoided in most instances.

The LUT400 pump control and level measurement system provides the optimum solution to manage collection networks. Pump status and pumped volume are monitored using the primary level device. Digital inputs capture site and pump data providing a completely integrated system.

Pumping stations are designed to handle specific flow rates – established by population served – but demographic changes alter requirements over time so design volume cannot be assumed. Pumped volume provides real trending data for collection system efficiency monitoring.

It is impossible to maintain completely sealed pipe work from pumping station to treatment works. Groundwater infiltration is inevitable. Pumped volume monitoring can provide network information to determine infiltration locations.

Network capacity can be effectively used by monitoring collection flows and transmitting data via telemetry to a SCADA system. This allows effective asset management and helps to diagnose structural problems.

Combined sewer overflows (CSO)

Combined sewage systems collect industrial, residential, and storm (rainwater) in the same sewer network. Under normal conditions, all of the sewage is piped to the sewage treatment facility for treatment and discharged back into the environment. However, during storm conditions, flows increase dramatically due to runoff and overflows often occur. When this happens, the sewage and rainwater overflows are discharged without treatment into the environment – stream, ponds, or rivers. These overflow events must be logged for regulatory compliance with the time, amount, and location recorded. Ultrasonic systems, like the LUT400, can monitor CSOs and record overflows which can be used for regulatory compliance.

Storm water ponds

In some cases where overflows often occur, storm water retention ponds are installed to collect the sewage and rainwater and prevent runoff to neighboring streams. Once the storm has subsided, the water is pumped back into the collection system for treatment. Ultrasonic systems are installed to monitor the level in these ponds and alert users to the presence of storm water and of possible overfilling of the pond.

Wastewater treatment plant

Main lift station

The main station is a large lift station located at the sewage treatment facility – all the remote lift stations within a city eventually pump their sewage to this central station. This application is similar to the remote lift station; however, the wetwell is much larger and there are generally fewer obstacles or obstructions. Ultrasonic systems monitor the level and control the pumps that convey the sewage into the plant as well as the sewage holdings and processes in the plant itself.

Overflow/storm tank

This overflow tank located at the sewage treatment facility provides the same function as storm water ponds. Acting as holding facilities, they collect the combined sewage and storm rainwater when the treatment plant is working at its maximum capacity. Once the incoming flow to the plant drops below maximum capacity, the overflow is pumped into the treatment works for processing.

Intake flow (influent flume)

Flow to the facility needs to be monitored and controlled to maintain plant efficiency. If inlet flow exceeds the rated capacity, it is diverted to storm holding tanks (see above) and later pumped back through the inlet once flows return to normal.

Inlet flow is often monitored using an open channel primary measuring device like a flume or weir and SITRANS LUT400, a secondary measuring device providing high accuracy level measurement using ultrasonic technology.

Flow data must be logged to ensure the following:

- that the treatment plant operates within tolerances
- compliance reporting for local environmental regulations is accurate

Bar screen

Inorganic debris and material is kept from entering the treatment plant by bar screens placed in the inlet channel. A buildup of debris on the upstream side causes a reduced level on the downstream side of the

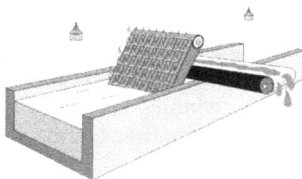

screen. When enough debris accumulates, a rake is activated to clear the screen and maintain flow. The debris is macerated and dried before disposal.

A Siemens HydroRanger 200 uses two ultrasonic transducers located at either side of the screen to measure the upstream and downstream levels. A relay onboard the HydroRanger 200 activates the rake at a user-defined differential level to clear debris. The use of non-contacting ultrasonic transducers ensures years of maintenance free operation.

Solids level measurement

Sand and debris removed by the rake and grit chambers is collected in a vessel (or a garbage dumpster) monitored by an ultrasonic system. Once full, the ultrasonic system provides an alarm to the user so that the vessel can be replaced with an empty one.

Chemical tanks

Chemicals, like the alum used as a flocculent in clarifiers or the chlorine that disinfects the water before final discharge, are stored on-site for daily use. Ultrasonic systems monitor the level in these vessels for inventory purposes.

Final effluent/discharge flow

All treatment plant discharges to the environment must conform to district legislation regarding quantity and quality, so accuracy is critical at this stage. Primary devices such as weirs or flumes through which the flow is channeled provide high accuracy measurement and are often used. The quality of the discharge must be monitored and is typically done with a sampler.

To achieve accurate flow monitoring and comply with legislation requirements, a primary measuring device combined with the LUT400 open channel monitor can be used. The LUT400 converts high accuracy ultrasonic level measurement to a flow value using known head versus flow formula or British and ISO Standards. Full data logging capabilities are available to log daily flow totals and flow rates.

Sludge treatment

The sludge that is left after the water has been removed is a toxic soup of waste products that must be treated very carefully. Its handling is highly regulated and all processes are carefully scrutinized. Measurement of inventory thus plays an important role in sludge management.

Digester gas

Biogases (methane) from the digester are often stored in a gas holder with a "floating" roof. These gases are re-used to provide heating for the plant or for local power generation. Ultrasonic systems are applied to monitor the height of the roof and thus the amount of gas present.

Digester level

This application was often done with ultrasonics with limited accuracy due the effects of methane on the speed of sound. Currently, radar devices like the SITRANS LR200 and Probe LR provide a more accurate, cost-effective reading and are replacing ultrasonics as the first choice technology for this application.

Dewatering/vacuum filter

The level of sludge in the vacuum filter is monitored by ultrasonics and controls a pump that will call for more sludge to the filter from the sludge tank when the level is too low.

Environmental applications

Application one: lift station control

A municipal district in Europe needed to upgrade its water treatment system and turned to Siemens to renew the local measurement devices, and control equipment and automation system. They also wanted to increase remote monitoring capacity.

Conditions

The project needed to integrate existing level measurement solutions such as hydrostatic pressure sensors with a new ultrasonic level measurement systems, lift pump control monitoring, and data acquisition.

- 116 pumping stations
- 18 associated communities
- population of 140,000

Solution

Siemens installed the following system:

- SITRANS LUT400 – the highly accurate ultrasonic controller from Siemens was chosen for the pump station monitoring and equipment control.
- Echomax XRS-5 ultrasonic transducers were installed to measure level.

With patented Sonic Intelligence echo-processing techniques, the LUT400 monitors and controls liquid levels in wells and/or liquid flow in flumes, weirs, or open channels. Its three relays control alarms or pumps The system monitors lift stations and collects data such as state of operation and pump run time. At each lift station, an LUT400 controller processes signals received from level measurement transducers.

Data is transmitted via landline using SITRANS RD500 to the control room at the district head office. The SITRANS RD500 has a built-in web server, FTP, and email client which allows the process to be monitored remotely. Alarm notifications are communicated through email and SMS text messages to one or more recipients to ensure that appropriate actions are taken by personnel.

Benefits

The SITRANS LUT400 proved a reliable and cost-effective solution for pump control, requiring less maintenance than the pressure sensors. SITRANS LUT400 increased efficiency and reduced maintenance costs.

Installing SITRANS LUT400 also optimized space in the electrical cabinet by rationalizing components and reducing wiring times. Existing equipment was easily integrated with LUT400 and it hooks seamlessly into the central control system.

Application two: wastewater plant control

A major Canadian city with a population exceeding 300,000 needed to upgrade its wastewater plants and pump stations.

All the locations were to link into a new high-speed data communications network, integrating them into a SCADA system. To ensure accurate data gathering, the city required reliable field instruments that could integrate into the network. After conducting a series of rigorous tests on ultrasonic level technologies from several suppliers, the city chose SITRANS LUT400 with its one millimeter accuracy and Siemens reliability.

Conditions
- six wastewater treatment plants handling a total of 208 million liters (55 million gallons) per day
- thirty-eight pumping stations with capacities ranging from 0.75 – 21 million liters (0.2 million to 21 million gallons) per day.
- twenty-five of the 38 are duplex pumping stations; the rest are complex pumping stations with huge variable frequency drive (VFD) pumps

Solution

Siemens installed the LUT400 in the 25 duplex pumping stations for continuous level monitoring and pump control and put Siemens Echomax transducers© into the wet wells.

The unique panel mount design of the SITRANS LUT400 made for quick installation, and for easy access. The controllers were networked into a PROFIBUS communication system. The SCADA system receives information from all stations, as well as other plant information, on the network. At the more complex pumping stations employing VFD pumps, the LUT400 measures level, and the analog output is fed into a PLC for pump control and remote telemetry.

Benefits

The ultrasonic systems perform reliably, and data collection from the many sites is now fast and accurate. Operators can monitor and control the plant sites from the central location. The automated system is more efficient, and has generated savings in energy and maintenance costs. Standardizing on LUT400 simplified installation and training, saving staff time and money.

Application three: small city operates at a high level

A small but growing Canadian city needed to increase its fresh water output due to a steadily growing population. This need was combined with a requirement to control costs and improve efficiency, all the while implementing new regulations requiring enhanced safety, training, and environmental protection.

Conditions

- Population of 74,000 and growing
- The system brings water from a river, cleans and treats it, then distributes it to residents
- Approximately 400 kilometers (250 miles) of water mains
- Pumps an average of 45 million liters (12 million US gallons) of water a day

Solution

The water plant embraces fully automated operations. All process points are integrated fully into a central SCADA system for continuous monitoring and control. Level monitoring is a critical need for a water plant, and Siemens installed more than 40 points of level in the main plant over the years. They monitor the river level, gate position in the river, as well as level in the flocculation tanks, sedimentation basins, settled water trough, filter beds, finished water storage reservoirs, chemical tanks, and other applications. An ultrasonic device monitors the floor as a safeguard to detect any spillage from a leak or ruptured pipe. At the city's various stations, reservoirs, and elevated water tanks, there are another 20 points of level being constantly monitored. Signals from remote sites are transmitted by radio link to the plant control center.

Benefits

The accurate and responsive Siemens sensors constantly monitor the river rate of change, a critical measurement. They also proved their preventative value when one of the interior tanks developed a leak and an alarm was immediately triggered, preventing a serious problem. Because Siemens has been the vendor of choice for quite some time, the plant actually has a number of older instrument models operating. These instruments still perform reliably and accurately, keeping maintenance costs down. The plant operators credit Siemens ultrasonic systems with helping them in their efforts to deliver quality water efficiently to city residents.

Application four: protecting water quality

A UK water company operates a large submerged microfiltration plant that services an entire city with potable water:

- The plant capacity is 84 megalitres a day
- Six filtration cells that remove particles as small as 0.2 microns, including bacteria, algae, and microorganisms.

Conditions

Two types of level monitoring is required:

- monitoring water levels is crucial because the flters require periodic backwashing with filtrate and air, and with chemicals when the maximum transmembrane pressure is reached. A filter blockage makes the water level rise, signaling the need for a backwash.
- monitoring the levels of the chemicals onsite, including sodium hypochlorite, sulphuric acid, sodium bisulphate, and sodium hydroxide (caustic soda). Reliable level measurement of the chemical storage tanks is crucial to successful process control, inventory management, and overflow prevention. Levels must also be measured in the clean-in-place tank (CIP) and the wash water tank.

Solution

Siemens Echomax XRS-5 ultrasonic transducers are installed on a "top hat" section of each chemical tank using a flange mounting. These transducers are hermetically sealed in chemically resistant enclosures

for reliable operation in harsh environments or chemical applications. The transducers are wired to the SITRANS LUT400 controllers which feed the 4 to 20 mA output directly into the SCADA control system.

Operators can now effectively monitor the level in each of the storage tanks. The ultrasonic systems work effectively without the complexity of higher priced solutions and without the problems of contacting and mechanical devices. Patented Sonic Intelligence advanced echo-processing technology is built-in to provide superior reliability.

Benefits

Non-contacting ultrasonic technology requires almost no maintenance, and eliminates worries about chemical compatibility. There is no drift from mechanical wear, and they are not affected by density changes so there is no need for recalibration.

They are easy to install and replace, with no need to drain the tank. Top mounting eliminates the need for isolation valves required for side-mounted pressure transmitters.

Food industry

Like many other industries, food applications can be divided into two types, storage and process. Furthermore, storage can be divided into solids and liquids. These applications are quite different and each presents its own unique, but not insurmountable, challenges for level measurement. For example, a 30 meter (100 ft) silo full of cornstarch is very different than a 1.5 meter (5 ft) stainless steel cooker vessel full of spaghetti sauce.

Solids

Solid storage applications are common; most food factories need to store a large supply of raw materials. Cereal milling companies, grain shipping, pet food, and breweries store raw grains (wheat, corn/maize, barley) in large tall silos or concrete bunkers; an XPS-30 mounted with an Easy Aimer II measuring grain at a brewing company is shown here.

Generally, longer range transducers are used as the applications are usually greater than 15 meters (50 ft) and the extra power of long range transducers overcomes attenuation due to dust (even on dusty applications that may be less than 15 meters (50 ft)). Easy Aimers are recommend so the transducers are mounted perpendicular to the material surfaces that often have a steep angle of repose. On excessively dusty applications, the SITRANS LR560 radar is the recommended instrument.

Often, many of the same type of vessels are found together, a cereal manufacturer storing a large inventory of grain, for example, and it is beneficial to use a multi-point system like the ten-point SITRANS LU10 to measure inventory of ten vessels with one controller.

Sanitary connection

Food processing vessels, whether blending fruit juices, cooking sauces, or mixing ice cream flavours, are much smaller, are made of stainless steel, and are cleaned regularly. A sanitary mounting process connection is required to ensure cleanliness. The Probe level transmitter and the XCT-8 transducer are available with a 4-inch sanitary clamp style process connection.

Temperature

The transducer has to be able to withstand the cleaning temperatures during the sanitizing process. In most cases the temperature is in the range of 80 °C and a mild caustic solution is sprayed on the inside of the tank. If the level device cannot withstand the cleaning temperature then it must be removed and sanitized separately; if the cleaning temperature is within the range of the transducer then it can be cleaned in place (often referred to as CIP).

Obstacles

The inside of a process tank has numerous obstacles that the level system has to take into account. The round ball is a cleaning ball and the sanitizing solution is sprayed into the tank from this ball. When mounting a transducer on a food process vessel, make sure that the cleaning ball is not within the beam path.

Liquids

Liquid storage vessels are the most straightforward measurements for ultrasonic systems in the food industry. The vessels are generally less than five meters (16.5 ft) in height and have little or no agitation. They contain food oils (vegetable oils), sugar syrups (glucose, molasses), cleaning solutions, water, acids, or fats.

Chemical industry

Applications with the chemical industry are generally limited to raw material storage vessels or finished goods storage – the process/reactor vessels are usually too hot or too high pressure for ultrasonic technology. However, other ancillary applications in the chemical industry require ultrasonic technology. For example, these may be the vessels storing lube oils for on-site machinery maintenance or containing the pre-treatment wastewater before flowing it into the municipal wastewater treatment system.

Storage vessels are typically bulk storage of liquid chemicals, hydrochloric acid, or sulfuric acid. Stored in vessels ten meters (33 ft) or less, they present no challenges to the ultrasonic system other than requiring that the wetted material of the transducer housing is compatible with the material in the vessel.

Ultrasonics is also used for inventory control on finished goods storage silos. A popular application is the containment of plastic pellets in silos. The SITRANS LR560 is also very suitable for this application.

The silos are tall and narrow and the pellets are pneumatically fed. Aiming is required because the pellets have a slight angle of repose and proper aiming increases the received echo strength and provides a more reliable signal.

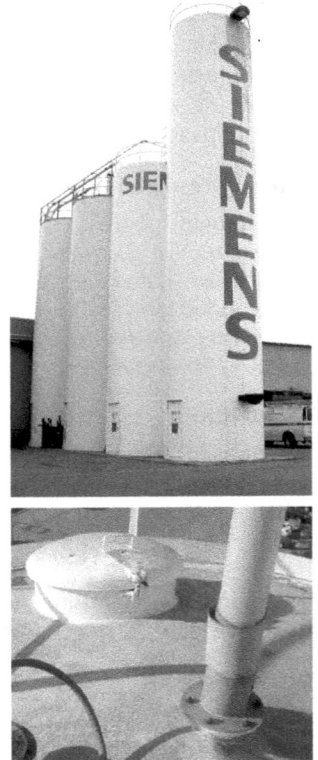

Application one: monitoring fluid level of drilling mud pits and tanks

A supplier of control and information systems for high quality oil and gas drilling equipment was looking for a technology to monitor fluid levels in drilling rig mud pits or tanks. Drilling mud is an essential component in drilling, and precise monitoring of the total mud volume in a rig's drilling fluid circulating system is essential for safe, efficient operation. Various technologies, including mechanical floats, had failed.

Drilling fluid or "mud" is pumped down through the drill pipe where it blows out through nozzles in the drill bit. The mud then flows back up the hole to the surface, clearing the hole by carrying the formation cuttings along with it. The mud lubricates the drill pipe and cools the drill bit. The weight of the mud column prevents formation fluids from entering the wellbore, preventing a "blowout." Through hydrostatic pressure, it also helps prevent caving. An increase in fluid levels may indicate gas, oil, or other fluids have entered the wellbore. If drilling fluid levels decrease, circulation is being lost to the formation. Left unchecked, either situation could result in a blowout.

Conditions

- Mud tanks are generally square or rectangular interconnected steel tanks used on offshore oil rigs to contain the large volumes of drilling mud flowing through the drilling fluid circulating system.
- On land rigs, these tanks are called mud pits, a throw back to the days when earthen pits were dug to contain drilling mud.

Solution

The company installed SITRANS Probe LUs. The non-contacting ultrasonic technology does not become fouled with material nor are there moving parts to wear out, making it practically maintenance free. The Probe LU's advanced echo processing ignores moving agitator blades and locks onto the targeted material level instead of obstructions. With its 2-inch NPT connection, it is simple to mount. It can be installed and calibrated regardless of material levels within a vessel. One unit works with all vessel sizes. Its range goes from less than 30 centimeters (one ft) to more than five meters (16 ft). There is an Intrinsically Safe version available for FM Class I, Division 1 applications.

Benefits

The company now has a reliable non-contact level measurement, avoiding electromechanical devices that are subject to wear and tear. The savings from reduced maintenance or replacement costs are significant.

Both safety and drilling efficiency are also increased as the company provides precise monitoring of a critical drilling variable.

Application two: inventory control

One of the largest by-products recovery firms in North America needed to measure level in their holding tanks. The tanks, indoor and outdoor, hold a variety of chemical and oil waste products ready for treatment. Government environmental regulations in such a facility are necessarily very stringent.

Before contacting Siemens, the company used floats with external sight board indicators. These mechanisms often stuck or froze, and the malfunctions made measurement unreliable. Plant personnel had to conduct frequent visual checks to verify that the readings were accurate, and to detect and fix equipment problems. Unreliable readings also made inventory management problematic.

Conditions

- Outdoor tanks for treating and storing solvents, waste oils, and caustics
- Indoor tanks for hazardous process water
- Mandatory high-level alarms to guard against spills

Solution

Siemens installed a complete ultrasonic level measurement system that offers both continuous level monitoring as well as high-point level detection. Each of the 19 tanks is equipped with an Echomax ultrasonic transducer, connected to one of three SITRANS LU10 transceivers. The system uses patented Sonic Intelligence software for enhanced reliability. Level data is fed to an HMI (Human-Machine Interface) located inside the testing lab. A SmartLinx interface card converts the signals to Modbus RTU protocol used by the HMI. Material level

data is routed to three SAM-20 devices that control the high-level alarm indicators, identify the tanks, and connect to a PLC that triggers both light and sound alarms if high levels occur.

Benefits

With automated, reliable level measurement in place, personnel no longer need to waste time with frequent manual inspections. Non-contacting ultrasonic technology requires almost no maintenance and, with no more float mechanical problems, the company has greatly reduced its maintenance costs.

Tank levels are monitored directly in real time. With exact content and storage data always available at the click of a button, the lab can schedule deliveries knowing storage space is available. Reliable continuous monitoring allows for effective scheduling and accurate inventory control. The high-point level alarms eliminate the risk of spillage, and help enhance plant safety.

Other industries

Application one: ultrasonic system for Peterborough Lift Lock

The Peterborough Lift Lock National Historic Site of Canada is one of 40 locks on the Trent-Severn Waterway, a national historic canal operated by Parks Canada.

Originally built for marine commerce, the waterway stretches 386 km from Lake Ontario's Bay of Quinte to Georgian Bay. Today, it is a world-famous tourist destination for recreational boaters and is regarded as one of the finest interconnected systems of navigation in the world. The historic lock stations of the Trent-Severn are central to numerous towns and villages in the waterway corridor. The Peterborough Lift Lock (150 km northeast of Toronto, Canada) is the world's highest hydraulic lift lock and an engineering marvel that attracts 160,000 visitors annually.

Conditions

The Lift Lock is a massive concrete structure supporting two chambers with rams connected in a closed water hydraulic system with a crossover valve. Operators flood the upper chamber with an extra

30 centimeters (12") of water, then open the crossover valve so that the heavy upper chamber pushes down, forcing the lower chamber up. Boats in the chambers are lifted or lowered a breathtaking 19.8 m (65 ft) in only minutes. Smooth operation depends on ensuring a differential of 30 centimeters of water. The weight differential of the extra water in the upper chamber drives the system that raises and lowers boats.

Throughout its history, the Lift Lock has relied on manual water level measurement but, when the Lift Lock celebrated its 100th anniversary in July 2004, it received a high tech birthday present from Siemens Milltronics who provided a new ultrasonic level measurement system. "The level measurement system is our birthday present to the Lift Lock as it turns 100 and our gift to the community as Siemens Milltronics celebrates its 50th anniversary this year," said Dave Bignell, President and CEO of Siemens Milltronics in 2004. "We're proud to have our technology installed in this landmark structure."

Solution

Echomax XRS-5 transducers are now mounted upstream and downstream in ABS tubes and enclosed in aluminum pipe. This protects the ultrasonic devices from being bumped or damaged by boats in the lock. The transducers are wired into a MultiRanger transceiver located in the control tower and it processes the signals and provides a digital readout of the water levels. MultiRanger is known for its reliable, accurate continuous level measurement. It's also compact, which is important in the tight quarters of the Lift Lock's small control tower.

Benefits

"The new system helps ensure proper operation of the lock," says Bruce Kitchen, Water Control Engineer for the Trent-Severn Waterway. "The ultrasonic level system enables the lockmaster to tell at a glance the elevations of the water upstream and downstream. Accurately monitoring and coordinating these levels help our staff achieve the optimum elevation differential for the working of this heritage machine."

Chapter Seven

Best in class –
the ultrasonic product line

You're simply the best, better than all the rest.[1]

The Siemens Milltronics line of sophisticated ultrasonic instruments is designed to meet level measurement challenges with accuracy, reliability, and longevity. The instruments represent the experience gathered from an excess of a million applications[2] and reflect the collected dedication and innovation of over a half a century of service by our enthusiastic and imaginative people.

From transistors to chips, from milliamps to PROFIBUS, from metal to PVDF, from a converted garage to a multinational company, Siemens Milltronics has pushed the boundaries of ultrasonic technology to take a leadership role in the water/wastewater industry on a global scale.

Our ultrasonic instruments meet the challenge of many types of applications, and the Siemens Milltronics catalog contains instruments for them all. In the following pages, we offer brief descriptions of both our latest and greatest, as well as the tried and true. Combined, they form a group ready to take on the measurement challenge, at any level.

SITRANS LUT400

SITRANS LUT400 is a high accuracy system that excels at continuous level monitoring and control in liquid, solid, or slurry applications in a wide range of industries.

---

[1] Tina Turner, *Simply the Best*. 1989 (originally done by Bonnie Tyler, 1988).
[2] Siemens Milltronics has installed over a million total applications, combining all technologies, including ultrasonics, radar, capacitance, and mechanical level measurement. All of this expertise is incorporated in every product designed and produced.

Three models make up the series:

- SITRANS LUT420 Level Controller
- SITRANS LUT430 Level, Pump, and Flow Controller
- SITRANS LUT440 High Accuracy Open Channel Monitor (OCM), which also provides a full suite of advanced level, volume, and pump controls

These controllers are a flexible solution for a world of applications: water/wastewater monitoring and pumping, inventory management, crusher control, truck load-outs, or anything in between.

The system is effective in wet wells, weirs, and flumes where build-up and turbulence are typical operating conditions. It can be customized to meet your specific application needs — from measuring flow rate in a narrow flume to volume in a ferric chloride storage bank.

In addition to monitoring flowrate in sewage works, SITRANS LUT400 can monitor industrial discharge and provide rainfall/storm water studies, inflow/infiltration studies, and sewer system evaluations. The programmable head versus flow curve accurately defines flow rate on unique or non-standard weirs and flumes.

The LUT400 has data logging and is adjustable, ranging from once per minute to once a day as it records the average flow rate for that time period. Daily recording includes min/max of temperature and flow rates, the time they occurred, and the daily total. Advanced functions include variable rate logging. It can be pre-programmed to log at a higher rate when needed.

The SITRANS LUT400 features a wall, pipe, or DIN rail mounted enclosure, housing electronics with a hermetically sealed, corrosion-resistant Echomax transducer. Components can be separated as far as 365 meters (1200 ft).

Key benefits

- industry-leading accuracy of ±1 mm (0.04") gives you confidence in your measurements
- easy to use – local user interface with four-button programming, menu-driven parameters, and Wizard support for key applications

- optional submergence shields to protect the transducer during flooding. Patented detection software can differentiate between a submerged condition and a high level.
- wall, pipe, and DIN rail mounting configurations with ¼-turn fasteners for quick access and hassle-free wiring with removable terminal strips
- outputs for alarms, chart recorders, controllers, and integration of existing systems
- energy saving functions with built-in real-time clock
- real-time clock control allows pump set points to alter according to the time of day. In a typical wet well installation, this could mean pumping out the well before peak rate and then deferring pumping during the expensive period. Penalties imposed by 'triads' and similar electric company incentives can be avoided automatically in most instances.
- special control mode to reduce grease rings and other deposits
- with a pump volume system using patented algorithms, the LUT400 continuously monitors inflow prior to a pump starting, thus calculating an accurate and timely inflow rate. When each pump starts, its current capacity is calculated and stored, giving accurate calculation of pumped volume throughput, unaffected by variable inflow. The unit does not require any discrete inputs from the pumps, or any form of in-line flow monitoring.

Key applications

- wet wells

- flumes/weirs

- chemical storage

- liquid storage

- bulk solids storage (gravel, flour bins, grains, cereals)

- plastic pellets

- crusher control

SITRANS Probe LU

SITRANS Probe LU is a 2-wire loop powered ultrasonic transmitter for level, volume, and flow monitoring of liquids in storage vessels, simple process vessels, and in open channels. SITRANS Probe LU is designed to monitor level in the water and wastewater industry and in chemical storage vessels.

SITRANS Probe LU features field-proven Sonic Intelligence signal processing that incorporates new echo processing capabilities and the latest micro-processor and communications technology, providing a continuous level measurement range of either six or twelve meters (20 or 40 feet), depending on application requirements.

SITRANS Probe LU occupies a unique place in the ultrasonic landscape. With its Auto False-Echo Suppression, it effectively eliminates vessel obstructions, negating any interference caused by extraneous vessel clutter. Furthermore, its advanced signal-to-noise ratio allows for clear echoes and a market-leading accuracy for transmitters of 0.15% of range.

Other application conditions like chemical contents and process temperature variations are easily dealt with because the transducer is available as ETFE[3] or PVDF[4], and an internal temperature sensor compensates for temperature changes.

Key benefits

- easy installation and simple startup
- programmable using infrared intrinsically-safe handheld programmer, SIMATIC PDM, or HART® Communicator
- communication using HART or PROFIBUS PA
- patented Sonic Intelligence signal processing
- extremely high signal-to-noise ratio
- Auto False-Echo Suppression for fixed obstruction avoidance

---

[3] Ethylene Tetrafluoroethylene
[4] Polyvinylidene Flouride

Key applications	
Storage vessels	Water industry

Storage vessels	Water industry
• acid	• charcoal filters
• lube oils	• flocculants
• juices	• storm water holding tanks
• vegetable oils	• chlorine storage
• resins	
• clean water towers	
• liquid sugars/corn syrup	
• fuel oils	

Sulfuric acid storage

Water storage

The Probe

The Probe is a short-range integrated ultrasonic level transmitter, ideal for liquids and slurries in open or closed vessels. The transducer is available in PVDF, making the device suitable for use in a wide variety of applications from water/wastewater to pizza sauce. It is ideal for the food and pharmaceutical industries as it can be quickly removed for cleaning.

The Probe is a workhorse, providing reliable level data based on Sonic Intelligence echo processing algorithms. It has an onboard filter that discriminates between true and false echoes from acoustical or electrical noises and agitator blades in motion, as well as temperature compensation. It also provides distance readings converted for display, analog output, and relay actuation.

Key benefits

- easy to install, program, and maintain
- accurate and reliable
- transducers available in PVDF
- sanitary model available
- patented Sonic Intelligence® echo-processing
- integral temperature compensation
- up to five meter (16.5 ft) measurement range

Key applications	
• chemical storage vessels	• mud pits
• filter beds	• liquid storage vessels

| Glue resins | Fruit juice | Drilling mud pits |

MultiRanger 100/200

MultiRanger is a versatile, short to medium-range ultrasonic single and multi-vessel level controller. Multi-Ranger offers true dual-point monitoring, digital communications with built-in Modbus® RTU via RS-485, as well as compatibility with SIMATIC PDM, allowing PC configuration and setup. MultiRanger contains Sonic

Intelligence advanced echo-processing software for increased reading reliability.

MultiRanger features the following:

- cost-effective level alarming
- ON/OFF and alternating pump control
- up to 15 meters (50 ft) measuring range
- monitors open channel flow
- converts level reading to volume
- measures differential level over a bar screen
- advanced relay alarming
- advanced pump control functions

MultiRanger is compatible with chemical-resistant Echomax transducers that can be used in hostile environments at temperatures as high as 145 °C (293 °F).

Key benefits

- digital input for back-up level override from point level device
- communication using built-in Modbus RTU via RS-485
- compatible with SmartLinx system and SIMATIC PDM configuration software
- single or dual-point level monitoring
- Auto False-Echo Suppression for fixed obstruction avoidance
- differential amplifier transceiver for common mode noise reduction and improved signal-to-noise ratio
- wall and panel mounting options

Key applications	
• water/wastewater	• woodchips
• fuel oil	• materials with high angles of repose
• municipal waste	
• acids	• bulk solids storage

MultiRanger monitoring
and storage vessel

Two MultiRangers measuring
ammonia tanks

MultiRanger measuring canal levels at
Peterborough Lift Lock, Ontario, Canada

SITRANS LU10

SITRANS LU10 is a long-range level monitoring system for liquids and solids, offering ten-point monitoring in a single unit for a wide range of applications to scan liquids, solids, or a combination of both contained in vessels of differing sizes, shapes, and configurations.

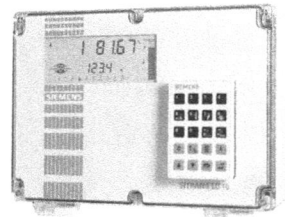

Using the patented Sonic Intelligence echo-processing software to measure level, space, distance, volume, or average/differential, the LU10 provides superior reliability and accurate measurement in vessels up to 60 meters in height (200 ft). Furthermore, transducers can be mounted up to 365 meters (1200 ft) from the monitor. Readings are displayed in user-selectable linear engineering units on the back-lit LCD.

Key benefits

- connection to a DCS or PLC using Siemens Milltronics Smart-Linx® interface modules for remote two-way communication and full parameter access[5]
- ten-point, long-range level monitoring
- automatic level-to-volume conversion for standard or custom tank shapes
- Dolphin Plus compatible
- easy to install, easy to program with removable infrared keypad
- ten mA outputs are available through external module SITRANS LU AO
- twenty relays for alarms are available through external module SITRANS LU SAM

Key applications

- chemical storage

- liquid storage

- bulk solids storage (sugar, flour bins, grains, cereals)

- plastic pellets

- tank farms

SITRANS LU10 monitoring coal bunkers

---

[5] Modules for popular industrial buses can be factory installed or added later to meet changing needs. No external gateway is required, reducing hardware and cabling costs.

HydroRanger 200

HydroRanger 200 is a level controller for up to six pumps, and provides control, differential control, and open channel flow monitoring.

For water authorities, municipal water, and wastewater plants, HydroRanger 200 is an economical, low-maintenance solution delivering control efficiency and productivity needed to meet today's exacting standards.

HydroRanger 200 uses proven continuous ultrasonic echo ranging technology to monitor water and wastewater of any consistency up to 15 meters (50 ft) in depth. Achievable resolution is 0.1% with accuracy to 0.25% of range. Unlike contacting devices, HydroRanger 200 is immune to problems caused by suspended solids, harsh corrosives, grease or silt in the effluent, reducing downtime.

Key benefits

- single or dual-point monitoring
- digital communications with built-in Modbus RTU via RS-485
- SIMATIC PDM compatible, allowing for PC configuration and set-up. Sonic Intelligence advanced echo-processing software provides increased reading reliability
- wall or panel mounting versions
- six relays on board as standard
- monitors wet wells, weirs, and flumes
- compatible with SmartLinx system
- anti-grease ring function
- differential amplifier transceiver for common mode noise rejection and improved signal-to-noise ratio
- MCERTS approved for open channel flow

Key applications		
• wet wells	• flumes/weirs	• bar screen control

Echomax Transducers

Echomax transducers fire ultrasonic pulses and measure the return echoes to determine the distance from the transducer to the material, whether it be a wide range of liquids or bulk solids applications. Designs are available with different measuring ranges, process connections, and ambient temperature specifications. The transducers are suitable for use in many applications from the benign clean water facilities to the aggressive chemical plants or dusty grain applications.

To select the proper transducer for the application, choose according to application requirements that include distance, temperature, and vessel configuration. The chart below indicates which transducers are suitable for the applications.

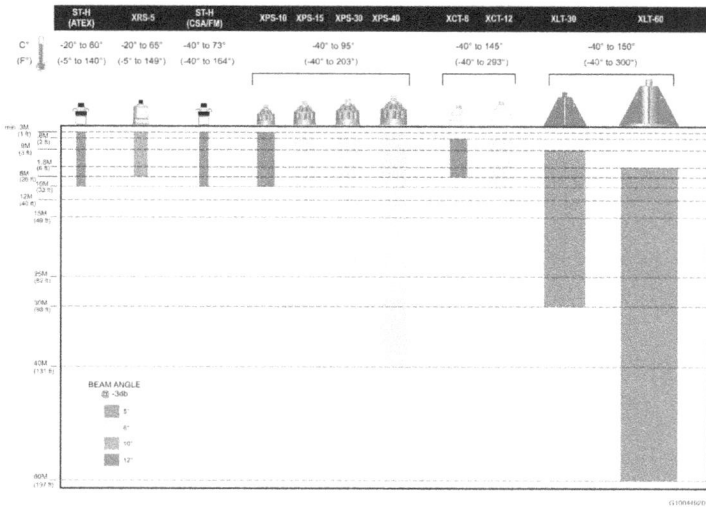

XRS-5

Echomax XRS-5 ultrasonic transducer provides reliable, continuous non-contacting level monitoring of liquids and slurries in narrow lift stations/wetwells, flumes, weirs, and filter beds. With a beam angle of just 10° and a CSM rubber face, it

has a measuring range from 0.3 to 8 meters (1 to 26 ft). Advanced echo processing ensures reliable data even in conditions with obstructions, turbulence, and foam.

The pre-molded, hermetically sealed rubber face and the PVDF enclosure are designed for maximum resistance to methane, salt water, caustics, and harsh chemicals common to wastewater installations. With an IP68 degree of protection, this rugged sensor is fully submersible in the event of flood conditions. If full submergence is possible in the application, use a submergence shield to maintain a high level reading output during submerged conditions.

Key benefits

- basic system for high/low alarm or dual pump control
- low-cost transducer compatible with a full range of Siemens Milltronics transceivers
- connects to advanced control systems with communications, telemetry, and SCADA integration capabilities
- chemically resistant PVDF copolymer enclosure and CSM rubber face
- measuring range: eight meters for measurement of liquids and slurries
- fully submersible: IP68 degree of protection
- easy installation with 1" NPT or 1" BSP connection

Monitoring level in sewage wet well

XPS/XCT Series

Echomax XPS/XCT transducers measure level in a wide range of liquids and solids. The transducers can be fully immersed, are resistant to steam and corrosive chemicals, and can be installed without flanges.

The XPS series offers versions for various measuring ranges up to 40 meters (130 ft), and up to a maximum temperature of 95 °C (203 °F). Two models have FM Class 1, Div. 1 approvals for applications with a measuring range up to fifteen meters (50 ft).

The XCT series can be used in applications at higher temperatures to measure level up to a distance of twelve meters (40 ft) and at a maximum temperature of 145 °C (293 °F).

During operation, Echomax transducers emit acoustic pulses in a narrow beam, so they are ideal transducers for dusty applications and for those with a lot of obstructions.

Key benefits

- integral temperature compensation
- low ringing effect reduces blanking distance
- self-cleaning and low maintenance
- connect using only two wires
- chemically resistant PVDF enclosure
- CSA, FM, ATEX and other global approvals
- hermetically sealed

XPS-15 measuring aggregate

XPS-30s measuring mining silos

XLT Series

Echomax XLT transducers operate in a wide range of bulk solids. Using a Siemens Milltronics controller, they measure from 0.9 to 60 meters (1.8 to 200 ft) in condition tempera-

tures up to 150 °C (300 °F). A beam angle of just 5° provides accurate readings in deep, narrow tanks.

With its highly developed signal sensitivity, the XLT transducer operates in a variety of difficult applications including limestone, cement clinker, and hot stone. All models have a sealed aluminum face to withstand harshest environments, and temperature variations are automatically compensated for by the integral temperature sensor.

Key benefits

- sealed aluminum face
- integral temperature sensor
- self-cleaning and low maintenance
- two-wire connection
- easy-to-install 1" threaded connection

ST-H

ST-H transducers measure level in chemical storage and liquid tanks. The narrow design of the ST-H allows the transducer to be mounted on a 2" standpipe. When mounted correctly, it is completely protected from the process, and can even be used in harsh, corrosive environments.

Key benefits

- can be mounted on a 2" standpipe
- immune to corrosive and harsh environments
- easy to install
- integral temperature sensor

ST-H monitoring a small ammonia tank

Conclusion

The Siemens Milltronics ultrasonic instruments offer customers a thorough level measurement solution, regardless of application. From small food tanks to ten-storey silos, ultrasonic measurement is reliable and cost-effective. It remains a vibrant measurement partner, along with a growing number of technologies offered by Siemens. Once the center of the industry, it is now joined by radar, frequency based capacitance, and TDR to offer a full complement of level application solutions. Level measurement in turn is an integral component of the Siemens Totally Integrated Solutions approach to customer requirements for process instruments, promoting a unified instrumentation strategy. Flow meters, valve positioners, temperature transmitters, pressure transmitters, and gas analyzers open up possibilities for application solutions other than those requiring ultrasonic level measurement.

This full armament of instruments and technologies is the strength of what Siemens can offer its customers, and ultrasonics beats strongly at its core. The technology has been hard at work for several decades and shows little sign of losing its popularity. Industrial developments like digital communications, trending, data logging, and software configuration packages keep it current and viable, alongside the whole of the instrumentation catalog.

Siemens Milltronics is determined to develop ultrasonic technology to its utmost capacity, and recent products like the SITRANS LUT400 illustrate the deployment of the technology to its maximum benefit by combining the best of the old with the best of the new.

The simplicity and precision of the SITRANS LUT400 exemplifies how ultrasonic technology is still a viable and vibrant player in the level measurement field. While other technologies like radar, capacitance, and TDR have their place, ultrasonics remains a cost-effective workhorse that offers reliable and long term service. When coupled with the latest advances in communications and software configuration, ultrasonic instrumentation remains made of sterner stuff and a significant player on the level measurement stage.

Index

Glossary

Entries in this glossary are from the *Operation of Municipal Wastewater Treatment Plants* text book, from the IEEE definition book, and from the contributions provided by Siemens Milltronics Process Instruments product and application specialists.

A

Accuracy	The absolute nearness to the truth. In physical measurements, it is the degree of agreement between the quantity measured and the actual quantity.
Aeration	1) The bringing about of intimate contact between air and a liquid by one or more of the following methods: a) spraying the liquid in the air b) bubbling air through the liquid c) agitating the liquid to promote surface absorption of air. 2) The supplying of the air to confined spaces under nappes, downstream from gates in conduits, and so on, to relieve low pressures and to replenish air entrained and removed from such confined spaces by flowing water. 3) Relief of the effects of cavitation by admitting air to the affected section.
Aeration Tank	A tank in which wastewater or other liquid is aerated.
Aerobic Digestion	The breakdown of suspended and dissolved organic matter in the presence of dissolved oxygen. An extension of the activated-sludge process, waste sludge is stored in an aerated tank where aerobic micro-organisms break down the material.
Agitator	Mechanical apparatus for mixing or aerating. A device for creating turbulence.
Algorithm	A prescribed set of well-defined rules or processes for the solution of a problem in a finite number of steps.
Allen-Bradley® RIO or AB® Rio	Allen-Bradley RIO protocol is an old protocol for communicating to remote I/O. It is only used by Allen-Bradley PLCs.
Alum, Aluminum	Used as a coagulant in filtration. Dissolved in water, it hydrolyses into $Al(OH)2$ Sulfate and sulfuric acid ($H2SO4$). To precipitate the hydroxide, as needed for [$Al2(SO4)3 \cdot 18H2O$] coagulation, the water must be alkaline.
Ambient	Generally refers to the prevailing dynamic environmental conditions in a given area.
Ambient Temperature	The temperature of the surrounding air that comes in contact with the equipment.
Anaerobic Digestion	The degradation of concentrated wastewater solids, during which anaerobic bacteria break down the organic material into inert solids, water, carbon dioxide, and methane.
Analog Signal	A signal where the value can be in a range of values (example: from 4 mA to 20 mA).
Antenna	An aerial which sends out and receives a signal in a specific direction. There are four basic types of antenna in radar level measurement. See *Horn, Parabolic, Rod,* and *Waveguide.*

Attenuation	A term used to denote a decrease in signal magnitude in transmission from one point to another. Attenuation may be expressed as a scalar ratio of the input magnitude to the output magnitude or in decibels.
Auto False-Echo Suppression	A technique used to adjust the level of a TVT curve to avoid the reading of false echoes. (See *TVT*).
Auto False-Echo Suppression Distance	Defines the endpoint of the TVT distance (See *TVT*). This is used in conjunction with auto false-echo suppression.
Automatic Sampling	Collecting of samples of prescribed volume over a defined time period by an apparatus designed to operate remotely without direct manual control. See *composite sample*.
Average Flow	Arithmetic average of flows measured at a given point.
Average Velocity	The average velocity of a stream flowing in a channel or conduit at a given cross section or in a given reach. It is equal to the discharge divided by the cross-sectional area of the section or the average cross-sectional area of the reach. Also called mean velocity.

B

Bar Screen	A screen composed of parallel bars, either vertical or inclined, placed in a waterway to catch debris. The screenings are raked from it either manually or automatically. Also called bar rack, rack.
Beam Angle	Angle diametrically subtended by the one-half power limits (-3 dB) of the sound beam.
Beam Spreading	The divergence of a beam as it travels through a medium. (http:// www.ndt.net/ article/az/ut_idx.htm)
Bias	In Siemens level equipment, there is a confidence bias value added in dB to the short shot confidence value. This is for comparison only. It allows the short distance to be chosen over the long shot distance with the same confidence value. Preset Milltronics equipment to 20 dB.
Biosolids	The organic product of municipal wastewater treatment that can be beneficially used.
Blanking	Zone extending downward from the transducer face in which it is received.
Bus	Network.
Bypass Pipe	A pipe that is mounted perpendicular to a vessel wall and is open to the vessel at the top and bottom. This is typically used on vessels that have a lot of turbulence or foam. The bypass pipe provides a calm liquid surface level equal to the level in the vessel.

C

Cavitation	1) The action, resulting from forcing a flow stream to change direction, in which reduced internal pressure causes dissolved gases to expand, creating negative pressure. Cavitation frequently causes pitting of the hydraulic structure affected.

2) The formation of a cavity between the downstream surface of a moving body (for example, the blade of a propeller) and a liquid normally in contact with it.

3) Describing the action of an operating centrifugal pump when it is attempting to discharge more water than suction can provide.

Centrifugal Pump A pump consisting of an impeller fixed on a rotating shaft and enclosed in a casing having an inlet and a discharge connection. The rotating impeller creates pressure in the liquid by the velocity derived from centrifugal force.

cfs (cu ft./sec.) The rate of flow of a material in cubic feet per second; used for measurement of water, wastewater, or gas; equals 2.832 x 10-2 m³/s.

Chlorination The application of chlorine or chlorine compounds to water or wastewater, generally for the purpose of disinfection, but frequently for chemical oxidation and odor control.

Chlorinator Any metering device used to add chlorine to water or wastewater.

Chlorine Contact Chamber A detention basin provided to diffuse chlorine through water or wastewater and to provide adequate contact time for disinfection. Also called a chlorination chamber or chlorination basin.

Clarification Any process or combination of processes whose primary purpose is to reduce the concentration of suspended matter in a liquid; formerly used as a synonym for settling or sedimentation. In recent years, the latter terms are preferable when describing settling processes.

Clarifier Any large circular or rectangular sedimentation tank used to remove settleable solids in water or wastewater. A special type of clarifier, called an upflow clarifier, uses flotation rather than sedimentation to remove solids.

Collection System In wastewater, a system of conduits, generally underground pipes, that receives and conveys sanitary wastewater or stormwater; in water supply, a system of conduits or canals used to capture a water supply and convey it to a common point.

Combined Sewer A sewer intended to receive both wastewater and storm or surface water.

CONF. CONF. is the confidence of signal value generated by the computer. It is based on all of the echo processing algorithms. It can be read in Run (key # 8) and program mode P805.

Confidence Describes the quality of an echo. The echo confidence is based on how well the transducer is mounted and aimed, and the noise floor associated with the echo profile. Transceivers display both the short and long echo confidence. If an echo is below a specific confidence threshold it is ignored by the Sonic Intelligence® software routines.

Confidence Threshold This is the minimum confidence value of an echo to be recognized as a valid echo (Parameter 805).

D

Daisy Chained Connected in series, one after the other. Cable is connected to the first device, then comes out of the first device and is connected to the second device and so on.

Damping	Term applied to the performance of an instrument to denote the manner in which the measurement settles to its steady indication after a change in the value of the level.
dB (decibel)	A unit used to measure the level of sound.
Dechlorination	The partial or complete reduction of residual chlorine by any chemical or physical process. Sulfur dioxide is frequently used for this purpose.
Defoamer	A material having low compatibility with foam and a low surface tension. Defoamers are used to control, prevent, or destroy various types of foam, the most widely used being silicone defoamers. A droplet of silicone defoamer contacting a bubble of foam will cause the bubble to undergo a local and drastic reduction in film strength, thereby breaking the film. Unchanged, the defoamer continues to contact other bubbles, thus breaking up the foam. A valuable property of most defoamers is their effectiveness in extremely low concentration. In addition to silicones, defoamers for special purposes are based on polyamides, vegetable oils, and stearic acid.
Device Description (DD)	A program file written in the HART or PROFIBUS Device Description Language (DDL) that contains an electronic description of all of a device's parameters and functions needed by a host application to communicate with the device.
DeviceNet™	A protocol developed by a group of companies as an open standard, used to establish master-slave/client-server communications between intelligent devices. It is based on the CAN (Control Area Network) protocol and is designed for connecting up intelligent devices such as limit switches and photo-electric sensors. See www.odva.org.
Dewater	To extract a portion of the water present in a sludge or slurry.
Differential Frequency	The frequency determined by calculating the difference between the transmitted and received frequency of the FMCW radar.
Digested Sludge	Sludge digested under either aerobic or anaerobic conditions until the volatile content has been reduced to the point at which the solids are relatively nonputrescible and inoffensive.
Digester	A tank or other vessel for the storage and anaerobic or aerobic decomposition of organic matter present in the sludge. See also *Aerobic Digestion*.
Digital Signal	A signal where the value can either be a logical 1 or a logical 0.
Dissolved Oxygen (DO)	The oxygen dissolved in liquid, usually expressed in milligrams per litre (mg/L) or percent saturation.
Domestic Wastewater	Wastewater derived principally from dwellings, business buildings, institutions and the link. It may or may not contain groundwater, surface water, or stormwater.
Doppler Beat Frequency	The frequency determined during the process of bouncing a continuous wave radar signal from a moving target.
"Dumb" Analog Input/Output Card	An input or output card that does not have HART communication protocol on it.

E

Echo	A signal that has been reflected with sufficient magnitude and delay to be perceived in some manner as a signal distinct from that directly transmitted. Echoes are frequently measured in decibels relative to the directly transmitted signal.
Echo Confidence	The recognition of the validity of the echo as industry level. A measure of echo reliability.
Echo Lock Window	A window centered on an echo in order to locate and display the echo's position and true reading.
Echo Marker	A marker that points to the processed echo.
Echo Processing	The process by which the radar unit determines the echo.
Echo Profile	A graphical display of a processed echo.
Echo Strength	Describes the strength of the selected echo in dB above 1 mV rms.
Electronic Device Description (EDD)	Another name for Device Description (DD). See definition for *Device Description*.
Electronic Device Descriptor Language (EDDL)	The language that the Device Description (DD) is written in.

F

False Echo	An echo caused by the internal structure of a container. Obstructions and incorrect unit placement will return echoes that do not reflect level.
Figure of Merit	A quantity characterizing performance of a device, system, or method in relation or comparison to similar alternatives. Engineering frequently defines figures of merit for materials or devices to determine their relative utility.
Float Switch	An electrical switch operated by a float in a tank or reservoir and usually controlling the motor of a pump.
Flocculant	Water-soluble organic polyelectrolytes that are used alone or in conjunction with inorganic coagulants, such as aluminum or iron salts, to agglomerate the solids present to form large, dense floc particles that settle rapidly.
Flocculating tank	A tank used for the formation of floc by the gentle agitation of liquid suspensions, with or without the aid of chemicals.
Flotation	1) Separation of suspended particles, or oil and grease, from solution by naturally or artificially raising them to the surface, usually with air. 2) Thickening of waste activated sludge by injecting air into it and introducing the mixture into a tank where the air buoys the sludge to the surface.
Flow	1) The movement of a stream of water or other fluid from place to place; the movement of silt, water, sand, or other material. 2) The fluid that is in motion. 3) The quantity or rate of movement of a fluid discharge; the total quantity carried by a stream.

4) To ensure forth or discharge.

5) The liquid or amount of liquid per unit time passing a given point.

Flow Rate The volume or mass of a gas, liquid, or solids material that passes through a cross section of conduit in a given time; measured in such units as kilograms per hour (kg/h), cubic metres per second (m3/s), litres per day (L/d), or gallons per day (gpd).

Flume A primary measuring device used to measure liquid flow (For example, see *Parshall flume*).

FMCW (Frequency modulated continuous wave) Indirect radar level measurement technique in which the radar device modulates the frequency of the transmitted signal. Level is measured by calculating the difference between the transmitted signal and the return signal.

Frequency The number of periods occurring per unit time. Frequency may be stated in cycles per second.

G

Gain Measures of the ability of a circuit to increase the power or amplitude of a signal from the input to the output.

Grit Chamber A detention chamber or an enlargement of a sewer designed to reduce the velocity of flow of the liquid to permit the separation of mineral (grit) from organic solids by differential sedimentation.

H

HART® Highway Addressable Remote Transducer. An open communication protocol used to address field instruments. HART is a registered trademark of HART Communication Foundation. See www.hartcomm.org for more details.

Head The height of the free surface of fluid above any point in a hydraulic system; a measure of the pressure or force exerted by the fluid.

Head Level The term used for the level of water in a weir.

I

Industrial Wastewater Wastewater derived from industrial sources or processes.

Influent Water, wastewater, or other liquid flowing into a reservoir, basin, treatment plant, or treatment process

Inlet 1) A surface connection to a drain pipe.

2) A structure a the diversion end of a conduit.

3) The upstream end of any structure through which water may flow.

4) A form of connection between the surface of the ground and a drain or sewer for the admission of surface or stormwater.

5) An intake.

Intermediate Frequency A frequency to which a signal wave is shifted locally as an intermediate step in transmission or reception.

Intrinsically Safe	Intrinsic Safety is a protection method, used for certifying electrical equipment to be used explosive atmospheres (flammable gas or combustible dust), in which any spark or thermal effect produced by the equipment is not capable of causing ignition.

J,K,L

Laminar Flow	Non-turbulent flow.
Long Shot	Describes a length of measurement beyond 1 m (3 ft.).

M

Manning formula	A formula for open-channel flow published by Robert Manning in 1890.
Margin	The difference calculated between the confidence and the threshold.
Master Device	A device in a master-slave system that initiates all transactions and commands (e.g. central controller).
Master Slave Protocol	Communication system in which all transactions are initiated by a master device and are received and responded to by a slave device.
Measurement	Each time a transmit pulse or set number of pulses is sent to the transducer.
Microwaves	The term for the electomagnetic frequencies occupying the portion of the radio frequency spectrum above 1 GHz.
Microwave Frequency	Microwave radar beams range in frequency between 5.8 GHz and 26 GHz.
Modbus®	A protocol used to establish master-slave/client-server communication between intelligent devices. It is a truly open, de facto standard, and the most widely used network protocol in the industrial manufacturing environment. Modbus is a registered trademark of Schneider Electric. See www.modbus.org.
Modem	A device that transmits data over a phone line by modulating the digital signals into analog signals at one side and transmitting that analog signal over a phone line. At the receiving end, it demodulates the analog signal back into a digital signal.
Motor Controller	A specialized type of controller whose typical functions performed by a motor controller include: starting, accelerating, stopping, reversing, and protecting motors.
Multi-master	More than one master device on the bus.
Multiple Echoes	Echoes caused by the parabolic effect of radar transmission inside a dished top vessel or a horizontal cylindrical tank. These types of echoes surpass the real level within a tank.
Multi-slave	More than one slave device on the bus.

N

Nappe	The sheet or curtain of water overflowing a weir or dam. When freely overflowing any given structure, it has a well-defined upper and lower surface.

Near Blanking	A blind zone that is created in front of a transducer or antenna to block the unit from reading false echoes.
Nozzle	A length of pipe mounted onto a vessel that supports the flange.

O

Open Channel	Any natural or artificial water conduit in which water flows with a free surface.
Open Channel Flow	Flow of a fluid with its surface exposed to the atmosphere. The conduit may be an open channel or a closed conduit flowing partly full.

P

Parameters	In programming, variables that are given constant values for specific purposes or processes.
Parshall Flume	A calibrated device developed by Ralph Parshall for measuring the flow of liquid in an open conduit consisting essentially of a contracting length, a throat, and an expanding length. At the throat is a sill over which the flow passes at Belanger's critical depth. The upper and lower heads are each measured at a definite distance from the sill. The lower head need not be measured unless the sill is submerged more than 67%.
Peer-to-Peer Protocol	Communication system in which all transactions can be initiated by any device on the network to any other device on the network.
Primary effluent	The liquid portion of wastewater leaving the primary treatment process.
Primary Sedimentation Tank	The first settling tank for the removal of settleable solids through which wastewater is passed in a treatment works. Sometimes called a primary clarifier.
Primary Sludge	Sludge obtained from a primary sedimentation tank.
Primary Treatment	1) The first major treatment in a wastewater treatment facility, used for the purpose of sedimentation. 2) The removal of a substantial amount of suspended matter, but little or no colloidal and dissolved matter. 3) Wastewater treatment processes usually consisting of clarification with or without chemical treatment to accomplish solid-liquid separation.
Process Device Manager (PDM)	A program made by Siemens that is used to configure, maintain, and troubleshoot field devices on HART and PROFIBUS networks.
Process Temperature	The temperature of the gas, vapour, or air located above the material being processed.
PROFIBUS	A communication protocol used to address field instruments. Different types include FMS (Fieldbus Messaging System), DP (Decentralized Periphery), PA (Process Automation). See www.profibus.com for more details.
Profile	See *"Echo Profile."*
Programmable Logic Controller (PLC)	An industrial computer that accepts inputs from the real world, runs a program, then generates outputs to the real world that is used to control processes.
Protocol	A set of rules that govern the exchange of information between two or more devices.

Pumping Station 1) A facility housing relatively large pumps and their appurtenances. Pump houses is the usual term for shelters for small water pumps.
2) A facility containing lift pumps to facilitate wastewater collection or reclaimed water distribution.

PTFE An acronym for Polytetrafluoroethylene.

Q, R

Radar Radar is an acronym for **RA**dio **D**etection **A**nd **R**anging. A device that radiates electromagnetic waves and uses the reflection of such waves from distant objects to determine their existence or position.

Range Distance between a transmitter and a target.

Range Extension The distance below the zero percent or empty point in a vessel.

Relay An electrical device that is designed to interpret input conditions in a prescribed manner and after a specified conditions are met, to respond to cause electrical operation or similar abrupt change in associated control circuits. The most common form of relay uses a coil and set of contacts. When current flows in the coil, contacts are opened or closed, depending on their arrangement. Relays are said to be normally closed.

Ringing The inherent nature of the transducer to continue vibrating after the transmit pulse has ceased. See *Blanking*.

S

Sampler A device used with or without flow measurement to obtain a portion of liquid for analytical purposes. May be designed for taking single samples (grab), composite samples, continuous samples, or periodic samples.

Scum 1) The extraneous or foreign matter that rises to the surface of a liquid and forms a layer or film there.
2) A residue deposited on a container or channel at the water surface.
3) A mass of solid matter that floats on the surface.

Secondary Effluent 1) The liquid portion of wastewater leaving secondary treatment.
2) An effluent that, with some exceptions, contains not more than 30 mg/L each (on a 30-day average basis) of BOD5 and suspended solids.

Secondary Sedimentation Tank A settling tank following secondary treatment designed to remove by gravity part of the suspended matter. Also called a secondary clarifier.

Secondary Treatment 1) Generally, a level of treatment that produces secondary effluent.
2) Sometimes used interchangeably with the concept of biological wastewater treatment, particularly the activated-sludge process. Commonly applied to treatment that consists chiefly of clarification followed by a biological process with separate sludge collection and handling.

Sedimentation 1) The process of subsidence and decomposition of suspended matter or other liquids by gravity. It is usually accomplished by reducing the velocity of the liquid below the point at which it can transport the suspended material. Also called settling. It may be enhanced by coagulation and flocculation.

2) Solid-liquid separation resulting from the application of an external force, usually settling in a clarifier under the force of gravity. It can be variously classed as discrete, flocculent, hindered, and zone sedimentation.

Sedimentation Tank
A basin or tank in which wastewater containing settleable solids is retained for removal of the suspended matter by gravity. Also called a sedimentation basin, settling basin, settling tank, or clarifier.

Settling Tank
A tank or basin in which water, wastewater, or other liquids containing settleable solids are retained for a sufficient time, and in which the velocity of flow is sufficiently low to remove by gravity a part of the suspended matter.

Short Shot
Describes a length of measurement within 1 m (3 ft.).

Shot
One transmit pulse or measurement.

Slave Device
A device (e.g. transmitter or valve) in a master-slave system that receives commands from a master device; a slave device cannot initiate a transaction (HART application guide, page 80).

Sludge
1) The accumulated solids separated from liquids during the treatment process that have not undergone a stabilization process.
2) The removed material resulting from chemical treatment, coagulation, flocculation, sedimentation, flotation, or biological oxidation of water or wastewater.
3) Any solid material containing large amounts of entrained water collected during water or wastewater treatment.

Sludge Blanket
Accumulation of sludge hydrodynamically suspended within and enclosed body of water or wastewater.

Slurry
A thick watery mud or any substance resembling it, such as lime slurry.

Smart Process Instrumentation
Microprocessor-based instrumentation that can be programmed. It has memory, and is capable of performing calculations and self-diagnostics and reporting faults, and can be communicated with from a remote location.

Stillpipe
A pipe that is mounted inside a vessel perpendicular to the vessel wall, and is open to the vessel at the bottom. This is typically used on vessels that have a lot of turbulence or foam. The stillpipe provides a calm liquid surface equal to the level in the vessel.

Submergence
1) The condition of a weir when the elevation of the water surface on the downstream side is equal to or higher than that of the weir crest.
2) The ratio, expressed as a percentage, of the height of the water surface downstream from a weir above the weir crest to the height of the water surface upstream above the weir crest. The distances upstream or downstream from the crest at which such elevations are measured are important, but have not been standardized.
3) In water power engineering, the ratio of tailwater elevation to the headwater elevation when both are higher than the crest. The overflow crest of the structure is the datum of reference. The distances upstream or downstream from the crest at which headwater and tailwater elevations are measured are important, but have not been standardized.
4) The depth of flooding over a pump suction inlet.

Suppressed Weir	A weir with one or both sides flush with the channel approach. This prevents contraction of the nappe adjacent to the flush side. The suppression may occur on one end or both ends.

T

Timed Relay	A relay whose opening or closing is delayed for a specified amount of time after the occurrence of a trigger event.
Totalizer	A device used to count pulses. Typically used to count the number of relay closures.
Totally Integrated Automation (TIA)	A Siemens concept for having all components of an automation system tied together using advanced communication protocols and totally integrated together.
Turbidity	1) A condition in water or wastewater caused by the presence of suspended matter and resulting in the scattering and absorption of light. 2) Any suspended solids imparting a visible haze or cloudiness to water that can be removed by filtration. 3) An analytical quantity usually reported in turbidity units determined by measurements of light scattering.
Turbulence	1) The fluid property that is characterized by irregular variation in the speed and direction of movement of individual particles or elements of the flow. 2) A state of flow of water in which the water is agitated by cross currents and eddies, as opposed to laminar, streamline, or viscous flow.
TVT (time varying threshold)	A time-varying curve that determines the threshold level at which the echo is processed.

U, V, W, X, Y, Z

Ullage	Amount of space left in a vessel in order for it to be filled. As opposed to the volume of material in a vessel.

Works Cited

Boyes, Wait. "The Changing State of the Art of Level measurement," *Flow Control Magazine,* February 1999.

Carlson, Al. "Open Channel Flowmetering," June 1998.

Carlson, Al. "The Right Transducer for the Job," June 1999.

Crabtree, Mick. *Pressure and Level,* Crown Publications (Pty) Ltd., 1998, pg 39–48.

Duncan, Doug. "Ultrasonic Sensors: Now an even better choice for solid material level detection," *Instrumentation & Control Systems.* November 1998.

Evans, Jack. "Ultrasonics comes of age for solids and liquids," *inTech.* April 1997, pg 39–42.

Felton, Bob. "Level measurement: Ancient Chore, modern tools," *InTech,* August 2001.

Grant, Douglas M. and Brian D. Dawson. Isco *Open Channel Flow Measurement Handbook.* Fifth Edition, ISCO Inc., 1997.

Hughes, Thomas A. *Level Measurement and Control,* ISA Press, 2002.

Li, Gordon. "Advanced Circuitry Improves Signal-to-Noise Ratio," 2003.

Massa, Frank. "Some personal recollections of early experiences on the new frontier of electroacoustics during the late 1920s and early 1930s," *The Journal of the Acoustical Society of America.* April 1985. pg 1296–1302.

Massa Products Corp. "Fundamentals of Electroacoustics,"

www.massa.com/fundamentals.htm

Milligan, Stephen. "How to Install Ultrasonic Transducers," 2003.

Sirlty, Paul A. "An Introduction to Ultrasonic Sensing," *Sensors,* Nov. 1989.

Check Out the Other Manufacturing Titles We Have!

Announcing Digital Content Crafted by Librarians